Trends and Debates in American Education

A Hispanic Perspective

Otto Federico von Feigenblatt

Trends and Debates in American Education

A Hispanic Perspective

Octaedro

Title: *Trends and Debates in American Education. A Hispanic Perspective*

First published: February 2023

© Ediciones OCTAEDRO, S.L.
C/ Bailén, 5 – 08010 Barcelona
Tel.: 93 246 40 02
octaedro@octaedro.com
www.octaedro.com

ISBN: 978-84-19690-06-7
Legal deposit: B 8209-2023
Design and production: Octaedro Editorial

Printing: Ulzama

Printed in Spain

This book is dedicated to the memory of my mother,
Norma Isabel, in gratitude for everything she taught me
and for having instilled the love of education in me.

Summary

Abstract

This volume covers a range of topics related to American education from a Hispanic point of view. Legislative intervention in the teaching of social studies in Florida, critical race theory in education, the strengths and weaknesses of the decentralized American education system, and the higher education of Hispanics in the United States, are some of the topics explored. The volume concludes with a critical interpretation of the shortage of teachers in the State of Florida at a time of great socioeconomic and political polarization in the United States of America, as a representative case of the central debates in education in the second decade of the twenty-first century.

Biography

Dr. Otto Federico von Feigenblatt holds a Bachelor's degree in social sciences from Ritsumeikan Asia Pacific University (Beppu, Japan), a Master's degree in International Development from Chulalongkorn University (Bangkok, Thailand), a Doctor of Philosophy in Conflict Analysis and Resolution from Nova Southeastern University, a Doctor of Education in Leadership in Education from Nova Southeastern University, and two postgraduate degrees in Social Justice and International Relations and a Master's degree in anthropology and archaeology from Harvard University (Boston, Massachusetts). Dr. von Feigenblatt has completed a postdoctoral fellowship at the University of San José, Costa Rica with a focus on management and leadership. From the research point of view, Dr. von Feigenblatt is the author of more than 100 academic articles published in indexed journals, several books, and has been a keynote speaker at important academic congresses internationally. With his administrative experience in the field of education and as a consultant in more than 20 countries, Dr. von Feigenblatt has received several international awards including being knighted by King Felipe VI of Spain with the Order of Isabella the Catholic, the Medal of the French Senate, the title of Honorary Ambassador of the Republic of Kosovo, the Dr. Vicente Rocafuerte Decoration from the Legislative Assembly of the Republic of Ecuador, among others. Dr. von Feigenblatt is a corresponding academician of the prestigious Royal Academy of Economic and Financial Sciences of Spain, an academician of honor of the Royal Academy of Doctors of Spain and is a numbered academician of the National Academy of History and Geography of Mexico. Britain's Brain Charity Trust chose him for the 2022 Brain of the Year Award for his contributions to the integration of mind maps in the field of education. He currently serves as Professor and Chair of the Graduate Education Department, Latin Division, of Keiser University (Fort Lauderdale, Florida).

Acknowledgements

This volume is the result of a great variety of experiences, re-search, and dialogues and therefore it is necessary to mention some of the key people and institutions to achieve its elaboration. First, I thank Keiser University, my academic home, and in particular the Latin Division, for serving as an institutional base and for the intellectually stimulating environment I have enjoyed. I also thank my doctoral students in educational leadership for the many stimulating dialogues on various topics covered in this book. We have learned a lot together and will continue to explore the frontiers of the discipline of educational leadership.

Interaction with great intellectuals has helped me develop some of the themes included in the book. I thank Dr. Olga Lucía Ostos Ortiz from the University of Santo Tomas in Colombia for her thoughts on the future of research in Latin American universities. I am also indebted to Dr. Óscar Yecid Aparicio Gómez of the International Journal of Theoretical and Practical Philosophy for sharing his thoughts on the need for interdisciplinarity in higher education. I also thank the rectors of the Complutense University (Madrid, Spain), University of Santo Tomas (Bogotá, Colombia), University of San José (San Carlos, Costa Rica, University of Yacambú (Venezuela), University of Guayaquil (Guayaquil, Ecuador), among others, for dedicating their valuable time to discuss issues related to higher education. I thank Dr. Theodore Macdonald of Harvard University for his guidance and sup- port in my growth as an anthropologist. I also thank him for the conversations about higher education in Latin America and for the motivation to integrate methods of anthropology applied to the field of education.

Finally, I am very grateful to my alma mater, Harvard University, for giving me the opportunity to grow as a researcher during the process of completing two graduate and master's degrees, and especially for the opportunity to conduct some of the research at the impressive Widener Library.

1

Legislative Intervention in Social Studies Education in Florida

Introduction

The U.S. education system has developed organically (von Feigenblatt, 2021c; Goldin, 1999). The decentralization of its development is reflected in its current structure. The Constitution of the United States has survived the test of time and although it has undergone modifications, its original structure endures (Gillion, 2020). One of the legacies of U.S. development is that powers that are not clearly enumerated in the constitution as powers of the federal government belong to the governments of individual states. One of the most important examples of these powers is power over education at all levels. Although the Federal government has a Department of Education, its role and authority differ greatly from its counterparts in countries with more centralized education systems (von Feigenblatt, 2021b; von Feigenblatt & Acuña, 2021; von Feigenblatt, Domíngue, & Vallé, 2015).

The third chapter of this book concentrates on the issue of the advantages and disadvantages of a decentralized system of education, but the brief explanation of the decentralization of the American educational system serves as a preamble to explore legislative interventions in education in the State of Florida. Known as the Sunshine State, Florida has a complex history. A Spanish possession until the nineteenth century, it is characterized by its ethnic diversity and by being an important tourist center (Fernandez-Armesto, 2003).

The sociocultural polarization afflicting the United States has affected all sectors of public policy (Chetty, Hendren, Jones, & Porter, 2018; Gillion, 2020; Kelin, 2020). Issues of education and in particular national identity have received increasing polit- ical attention at all levels. Florida's unique characteristics, such as a large Hispanic immigrant population and a large rural area, have turned the southern state into a battleground on issues of national interest such as the social science and civil rights curriculum (Osorio, 2021; Shumow & Pinto, 2014).

The State of Florida has one of the lowest unemployment rates in the entire country, 3.3% according to March 2022 data («Latest Statistics», 2022). Although one of the state's main industries is tourism and services, the state governments pro-free market policies have managed to facilitate a relatively rapid recovery. The fiscal facilities of the State and the conjuncture of a severe economic and political crisis in several neighboring countries have had the positive externality of promoting the in- vestment of a large number of foreign and American entrepreneurs («Florida Realtors Market Data», 2022). Among the effects that the flow of extra-state investment has had has been a boom in the real estate sector. The economic growth of the State has resulted in significant socio-cultural and demographic changes.

Florida is not the only American state that has received a large influx of immigrants, but immigration to Florida has several characteristics that have had a great influence on socio-political trends in the Sunshine State. An ethno-national group with great influence in South Florida is the Cuban community (Gonzalez, 2010; Sabogal, 2012; Shumow & Pinto, 2014). Important since before the great influence caused by the communist revolution on the island. The Cuban community in Florida has achieved great success in the private and public sectors (Osorio, 2021; Pentón, 2021). Proximity to the island is an important factor but the political situation in Cuba has resulted in a marked emphasis on anti-communism and anti-socialism. Among the many influential Cubans in Florida, we can mention Senator Marco Rubio, the Lieutenant Governor of Florida, Ms. Núñez, the Mayor of the City of Miami, Francis Suarez, among others (Osorio, 2021; Shumow & Pinto, 2014).

Another important group in South Florida is made up of Venezuelans who have become immigrants because of the deep political crisis in their country. Among Venezuela's immigrants are some of the Country's leading families, and they have invested in businesses and real estate in cities such as Weston and Doral (Sabogal, 2012). Venezuelan immigrants also tend to be conservative and anti-socialist because of the political crisis in their country and also because many represent the transplanted elites of their homeland. Other ethnic groups present in Florida include Haitians, Nicaraguans, Vietnamese, and Hondurans, among others. Although it is difficult to generalize about the characteristics of the immigrants from these countries, studies indicate that there is a great polarization from the point of view of socio-economic status of newcomers (Jimenez, 2017). From former hierarchs and big businessmen to illiterate peasants. One of the few things in common between the two socio-economic extremes is that they share a conservative view from the point of view of community values (Gillion, 2020).

Thus, although they have very different opportunities in Florida and live in parallel worlds, they share a conservative vision of family and education (Gonzalez, 2010). North Florida is an agricultural region, historically linked to the legacy of the old aristocratic South before the civil war. Rural areas in the United States tend to be conservative and are regularly political strongholds of the Republican Party. Florida is no exception. A political map of Florida shows that the strength of the Democratic Party is concentrated in urban centres in central and northern Florida while South Florida, due to the large concentration of Hispanics tends to be Republican (Kelin, 2020). The result of demographic characteristics, history, and economic development has been a conservative tendency in aspects of social values and a neoliberal tendency from the economic point of view.

Another unique feature of the State of Florida is the high number of Northern pensioners residing in the Sunshine State. Many of those pensioners come from northern states with much more liberal tendencies and because of their high purchasing power, they settle in upscale urban regions, such as Broward and Palm Beach counties in northeastern South Florida. These en- claves have a distinct socio-political culture from the rest of Florida and due to the decentralized school system of the United States; the sociocultural difference is reflected in the composition of the school boards of the districts located in those enclaves. The result of the socio-political characteristics previously exposed has been a socio-political polarization between the Florida Department of Education and the various school districts, which reflect the demographic composition and socio-political trends of the regions where it is located. Polarization is reflected in moments of crisis and in debates on issues related to traditional values. At the same time, the faculty also does not reflect the demographic characteristics of the parents and students that make up their school districts. The nature of the teaching career and the requirements to obtain teacher certification in the State of Florida results in a group with sociocultural characteristics that are not very representative of the population they serve. Teachers must have at least a bachelor's degree from a recognized university, one year of supervised experience as a teacher's aide, and pass three statewide exams in order to receive professional certification. Many teachers come from other states and by the nature of their profession tend to have a higher level of education than the general population.

For all the above, Florida has become a battleground in education. The centrality of education-related issues is evident in the attention given to education by the governor and legislature («Civic Education Curriculum», 2021; «Individual Freedom», 2022; «Parental Rights in Education», 2022). Governor DeSantis in collaboration with the Republican majority in the Florida legislature has passed two laws directly aimed at public education («Individual Freedom», 2022; «Parental Rights in Education», 2022). One of these is entitled the Fatherhood Rights in Education Act and the other is the Individual Liberty Act. The next section of the chapter explores the importance of each of the laws and the potential impact they can have on social studies education.

CS/CS/HB 1557 Parental Rights in Education Act

One of the most controversial issues in Florida is gender and in particular sex education in public schools and colleges. The great polarization between the values of much of the population in rural areas compared to some urban areas such as the city of Orlando and Miami Beach. Because some Florida cities are national pilgrimage centers for nightlife. The City of Orlando and South Beach in particular have become important recreation centers for LGBT communities. Due to the geographical separation between urban and rural areas, the existence of such centers has not become the central issue of discord between the various socio-cultural groups, but the inclusion of gender issues in the school curriculum, particularly in the primary years, has spilled the proverbial «glass» in the Sunshine State. The inclusion of books in school curricula and libraries on gender issues written for early developmental ages has generated a strong reaction of rejection from groups of parents and legislators. In particular, the inclusion of some children's books with controversial content on gender and sexuality issues without informing parents has caused a loss of confidence in school boards and in particular in education officials. The perception that many decisions about the values taught in classrooms, part of the hidden curriculum, are made without any consideration of parents' opinions or preferences has resulted in interest groups asking the governor and legislature to intervene directly to protect the rights of parents in their children's education.

The result has been the Parental Rights in Education Act, known to critics as the «Don't Say Gay Bill» («Parental Rights in Education» Act, 2022). This law limits sex education at an early age and regulates it for the more advanced grades. The law sim- ply reiterates the need for districts and teachers to focus on teach- ing educational standards and not take the liberty of including a «hidden curriculum». The law also authorizes the Florida Department of Education to intervene if any district includes gen- der instruction that does not adhere to the guidelines set forth in the standards and recommendations. Another peculiarity of the law is that it allows parents and other stakeholders to directly sue districts in cases where the law is not followed. One of the main challenges for the Department of Education and districts has been to monitor what happens in classrooms. Because of respect for academic freedom and the absence of video cameras in classrooms, it is very difficult for district officials to find out what is being talked about in classrooms.

Most of the information on the topics included in the hidden curriculum comes from student testimonials and in some cases from cell phone recordings. Still, the only recourse parents have had for this kind of academic over-discretion has been to inform district officials, who also typically give the principals of each institution considerable discretion in dealing with this type of matter. Therefore, in a school or college where the principal and teachers believe that it is right to include a hidden curriculum on gender diversity issues, it is very difficult for parents to take effective measures to ensure that standards are respected, and they are informed about the content of the lessons.

The Parental Rights in Education Act went into effect on July 1, 2022, and includes direct mechanisms for monitoring and implementing the law through direct community monitoring («Parental Rights in Education», 2022). The Act also empowers parents directly to request information about cases where there is any suspicion and also to request access to the supporting texts of each lesson. Thus, the Act increases transparency in the classroom and allows parents to evade school bureaucracy to obtain certain types of information about their children's education. The main mechanism of implementation is the existing state court system since any interested party can sue and thus force a possible investigation.

There are many opponents of the law. Major professional teachers' associations have opposed the law because they consider it unnecessary and overly ambiguous. Another important criticism is that it limits academic freedom and invites abuses in monitoring. Many teachers allege that cases of abuse of academic freedom are very rare and that therefore the law is not necessary and is onerous because it adds an extra level of evaluation for teachers.

There are also detractors for ideological reasons. Nonprofits and corporations aligned to a liberal agenda in aspects of social values, reject restrictions due to missed opportunities to socialize youth with values compatible with their ideological agendas. Examples include, *inter alia*, the Disney Corporation, and a wide range of LBGT community support organizations (Whitten, 2022). In this case, the rejection of the law is because they consider that the law prevents them from including a hidden curriculum that gradually instills in children an alternative vision of gender and sexuality with the goal of normalizing gender diversity. In other words, this group of NGOs, activists, and corporations want to have the discretion to be able to include the hidden curriculum and thus carry out a slow but sure sociocultural change.

Groups supporting the law include, *inter alia*, conservative nongovernmental organizations, religious groups, parent groups, and the Republican Party. The reasons why these groups support the law vary. Religious groups tend to defend the Jewish-Christian view of family and sex, and therefore oppose a hidden curriculum that rejects their view of the family and intrudes on the education of values and morals of children and adolescents (Kamenetz, 2021). Other conservative groups oppose it be- cause of the principle of separation of duties. In the American tradition of public education, education about values and morals is relegated to parents. Interference in matters of fundamental values is perceived as an abuse of the freedom of children and parents to make their own choices (von Feigenblatt, 2021a, 2021b; Goldin, 1999). Ironically, part of the argument is the separation of church and school that exists in the United States. In principle, students in public schools cannot be indoctrinated with religious philosophies regardless of the personal preferences of teachers or principals.

Therefore, since churches and religious teachers refrain from talking about religion in classrooms, out of respect for the principle of separation, teachers, and socio-cultural groups with alternative values are also expected to refrain from inculcating those values in pupils in school classrooms. In some cases, the argument of financing the public school system is also used (Goldin, 1999; Lee & Jimenez, 2021). A large portion of school districts' budget comes from local property taxes. Due to the high cost of real estate in much of Florida, and the difficulties in obtaining bank credit to acquire a property for young people, most owners tend to be elderly or people of high purchasing power (Sabogal, 2012).

These two groups tend to be more conservative, have fewer children, and in many cases send their children to private or religious schools (Chetty *et al.*, 2018; Gillion, 2020). In these cases, the owners argue that not only do they not directly benefit from their property taxes by not using the services, but that their contributions are being used to promote an ideological agenda that they reject. An alternative that has been presented to the legislature is the possibility for owners to receive vouchers equivalent to their contribution to public education so that they can use them in private colleges and schools (Gillion, 2020; Jimenez, 2017). This option would be disastrous for public education and therefore the legislature has tried to look for ways to not reach that point. The law under discussion is a way of arguing that public education is not being used by certain groups for the ideological indoctrination of children and therefore continues to be a public good and a good investment.

The possibility of vouchers for private education is a very tangible consideration for the public education system, because there is already a voucher system for students with special needs (Chetty *et al.*, 2018). Parents of students with special needs, disabilities, can apply for the equivalent of $10,000 annually to help pay for tuition at a private institution. Although some parents do not take advantage of this opportunity because many private institutions have a wide variety of extra costs beyond schooling, including uniforms, books, etc., etc., etc. But there are also middle-class parents who try to identify minor special needs for their children as an excuse to apply for a voucher to pay for part of the schooling in a private institution.

This type of coupon is called the McKay Scholarship program, 2022. Some private institutions try to keep the cost of their schooling very close to $10,000 annually to attract this type of student and add some extra costs to filter students of a low socioeconomic status. Therefore, it is interesting that most private primary and secondary institutions are distributed from the point of view of the cost of schooling in the $10,000 or $20,000 range. The strategic goal of these institutions is to divide middle-class students with up- per-class students. The reason this practice is considered a goal is the strong correlation in the United States between socioeccmic status and academic performance. For institutions with tuition fees of over $20,000, most do not accept McKay Grants.

From the point of view of many parents from middle-class families, they consider McKay scholarships as their own money they have paid in property taxes. There is also a strong relation- ship between this practice and religious affiliation (Gillion, 2020). The separation of church and state in the field of education. Some religious families use these vouchers as a way to register their children in private religious schools. Even so, current school voucher legislation, the case of McKay Scholarships, requires the child to be evaluated by an expert to certify that they have special needs («McKay Scholarship», 2022). Although the field of special education in the United States has expanded the definition of the term in recent years, it is not possible for parents to get special education experts to find any justification for certifying that they have special needs. The irony is that the McKay scholarships do not apply to gifted children, and therefore many middle-class parents with gifted children in public institutions have tried to find reasons to apply for vouchers so they can send their children to private institutions.

To understand the importance of the Parental Rights in Education Act, it is necessary to consider the debate on private school vouchers. If we interpret the Law from this point of view, the law is a compromise between various interest groups and reflects the great power that teachers' associations have. While the law limits teachers' and districts' discretion in curriculum content, it represents a victory in protecting public education funding and presents a useful tool to deter middle-class parents who are considering enrolling their children in religious schools. Considering the demographic characteristics of Florida, the Act is a Solomonic solution to a crisis in primary and secondary education, which reaffirms the basic principles of the American educational system.

CS/HB 7 Individual Freedom in Education Act

The Individual Freedom Act also went into effect on July 1, 2022 («Individual Freedom», 2022). This law was passed by the Florida legislature and signed by Governor Ron DeSantis in a sociocultural and political context similar to the Fatherhood Rights Act. One difference is the emphasis of the Law. The Parental Rights in Education Act focuses on protecting the curriculum and instruction from the introduction of alternative gender and sexuality ideologies, while the Individual Freedom in Education Act focuses on protecting individual identity from harmful racial stereotypes («Individual Freedom», 2022; «Parental Rights in Education», 2022).

It is difficult to understand the raison d'être of this law without taking into account the socio-political context of the United States. In July 2013, a movement started called «Black Lives Matter» to fight perceived police abuses of African-Americans (Bloch, 2021). The trigger for the establishment of the movement was the death of Treyvon Martin, an African- American boy who died at the hands of a volunteer guard in Florida. The ethnicity of the guard was an important point of contention at the socio-political level. George Zimmerman, a young Hispanic man of white complexion, during an altercation in a private residential community shot Martin in self-defense. In a nationwide trial because of its symbolism, an ethnically-racially diverse jury found Zimmerman not guilty of all charges. Florida has a law called «Stand your Ground» that allows the defense of private property with force if necessary («Justifiable Use of Force», 2021). Although the case mostly focused on matters of private property and self-defense, radical groups at the national level identified an excellent opportunity to introduce aspects of racial justice into the public sphere.

Another trigger for the legislature and parent groups was the introduction of the famous 1619 curriculum («The 1619 Project: Education Materials Collection», 2022). This curriculum began as a journalistic project of the New York Times and its first part was published in 2019. Journalist Nikole Hannah-Jones developed a project to celebrate the 400th anniversary of the arrival of the first slaves in the United States («The 1619 Project: Education Materials Collection», 2022). The project was controversial from the beginning due to the historical errors included in it. Another controversial aspect was the integration of the so-called Critical Racial Theory. This theory is not new and derives from the traditional critical theory of Marxist roots (Cotton, 2009; Tang, 2008). Critical Race Theory basically alleges that the socio-political institutions of the United States are essentially racist and that the war of independence was waged to protect the institution from slavery (Hilburn, Journell, & Buchanan, 2016). Beyond the assertion previously exposed, the project also identifies 1619 as the true founding year for the nation. The project received extensive media coverage and was adopted in several school districts in regions controlled by liberal groups.

On the other hand, the 1619 project has received a lot of criticism from historians (Kaufman, 2019). The criticisms of several prominent historians did not focus on issues of critical racial theory, but on the historical errors included in the project. Another group of critics focused on aspects of structural violence and the inclusion of activities in the project focused on engendering feelings of guilt among white children and adolescents. A central idea of the project is that there is a collective guilt and in parallel there is also a feeling of group victimization. One of the controversial parts of the project is the implementation of lessons for primary and secondary school students. In some activities the emphasis is on white students feeling guilty about slavery. Another aspect that has also received criticism is the conclusion or recommendation that collective compensation be paid to African Americans for the damages caused during the period of slavery.

A central tenet in U.S. culture and social development is the principle of individual rights and liberty (Gillion, 2020). The idea of hereditary and collective guilt is anathema to a large group of Americans. Beyond deep discrepancies over the interpretation of the country's formative years, imposing collective guilt on children has caused a strong reaction of rejection and anger from groups of parents and various political groups. In other words, opposition to the recommendations and implementation of the curriculum developed as part of the 1612 Project includes two fundamental principles. The first is the discrepancy over the interpretation of the motivations of the country's founding fathers. The second is respect for individual freedom and the rejection of any kind of hereditary status.

The previous discussion on the two central principles of the discrepancy, explain the content of the Individual Freedom in Education Act and its name. The emphasis of the law is to protect primary and secondary students from curricular changes and prescribed texts, which engender feelings of collective guilt. The law simply reinforces existing mechanisms for monitoring and controlling the implementation of the social science curriculum to prevent cases of teachers integrating critical racial theory and related projects into the so-called hidden curriculum. The Individual Freedom in Education Act is not only a reactive law but also includes provisions to cooperate with districts, through the Florida Department of Education, in developing appropriate standards and teaching materials in the social sciences.

An important provision of the Individual Liberty Act is the training and regulation of teacher professional training activities and programs. The ambiguity of some of the standards in teach- ing U.S. history and world history, combined with the great discretion and academic freedom of teachers, exacerbated by the lack of training of some teachers, has resulted in gross abuses on the part of some ideologically radical teachers (von Feigenblatt & Acuña, 2021). Historically, the function of academic freedom in the United States has been to allow teachers and professors to explore potentially controversial issues in a context of constructive dialogue (Goldin, 1999). The exams for professional teacher certification in the State of Florida clearly mention that teachers should not share their political opinions in the classroom and that they should present sensitive issues in a balanced way and include different perspectives. Unfortunately, there have been cases of teachers who have openly rejected the academic neutrality norm and have actively promoted political movements in their classrooms (Bloch, 2021).

A prominent example of the phenomenon previously exposed in the State of Florida, is the controversial case of Amy Donofrio in Duval County (Bloch, 2021). The middle-grade teacher includes in her lessons a lot of comments about the radical movement called Black Lives Matter. However, the teacher did not just make a few comments in favor of the group, but she also dressed in T-shirts with phrases of support for the group. The Florida Commissioner of Education even accused the teacher of indoctrinating students (Bloch, 2021). After multiple warnings from her superiors, the teacher continued her classroom activities and was eventually suspended from her job and her professional certification was also suspended.

The teacher sued the school district requesting that her position be reinstated, and she be paid damages. After negotiation between the teacher's legal team and the district's legal team, an agreement was reached outside the court that she had to sign, which meant that she will never be able to be a teacher in the county again and that she will receive a sum of money as compensation (Bloch, 2021). Because the courts did not rule on the case, the possible outcome cannot be predicted with certainty. What is known is that Donofrio reported that although she had no better alternative than to accept the agreement, she was not deeply saddened by the clause on the prohibition of teaching in the County (Bloch, 2021).

It is necessary to understand the American legal system to assess the importance of this case. Donofrio's legal team, with great help from ideologically aligned groups, realized that in the long run she was going to lose the case and that a bad deal was better than a long case with little chance of victory. The inclusion of the clause on teaching and its acceptance by Donofrio is an important indication of the weakness of the case. The principle of precedent is central to the American legal system (Benlevy, 2000). One of the reasons why national lobbyists support this type of case at the local level is because of the symbolism and importance as a legal precedent. One case that received national attention ended up with a suboptimal outcome from the teacher's point of view and one totally negative from the point of view of her many support groups. The possibility of appealing a possible negative result was also not feasible due to the conservative majority in the Supreme Court.

The combined effect of cases such as that of teacher Donofrio, the statements of the Governor and the Commissioner of Education, and the Individual Freedom in Education Act, can potentially be a deterrent from the point of view of self-selection in a teaching career. Notably, debates over the aforementioned laws in the State of Florida have occurred during the crisis of teacher shortages and vigorous efforts at the state level to raise the minimum wage for new teachers. The discussions have been concentrated on efforts to raise the salary of novice teachers and not on increasing the salaries of veteran teachers (Trischler, 2022). One possible interpretation of the Individual Freedom in Education Act in conjunction with the context of state-level wage increases is a human resource renewal strategy. The incentive of higher initial salaries, $47,000 plus benefits, combined with a legal framework limiting teachers' political activism, alternative routes to enter the teaching career for professionals in other disciplines, gives little incentive for veteran teachers to continue teaching.

Also, the COVID-19 crisis has increased the retirement rate of teachers and the aforementioned incentives have increased the number of professionals who have not studied in educational faculties who are considering seeking a teaching career (Trischler, 2022). Thus, it is possible to interpret the Individual Freedom in Education Act in context as an effort to attract young professionals with less radical ideological profiles. Another way of looking at it is that conditions are being fostered to renew a faculty perceived by the State government as highly radicalized ideologically with a more docile and technocratic faculty.

In short, the Individual Freedom in Education Act responds to a legitimate concern of parents and community leaders about the content of youth education in the public education system. The concept of academic freedom is also at stake because of the abuse of teaching discretion by some radicalized teachers. Finally, a concerted strategy at the state level of transforming faculty into an a-political and technocratic direction can also be elucidated.

Conclusions

Legislative intervention in education is not a new phenomenon in continental Europe and Latin America, but it is in the State of Florida and in the United States in general. A history of local development of public education, combined with a strong principle of academic freedom and self-regulation in the education sector has been undermined by the pernicious combination of a highly polarized socio-political environment and isolated cases of abuse by radicalized teachers of their discretion in the classroom. The legislative intervention in primary and secondary education represented by laws such as the Parental Rights in Education Act and the Individual Freedom in Education Act are symptoms that homeostasis in the system of self-regulation in public education is in crisis. For now, legislative interventions have been relatively benign, avoiding excessive government regulation of education and rather serving as incentives to restore the political neutrality of teaching.

However, after having broken a historical legislative taboo in relation to public instruction, it is possible that the possible insubordination of some elements in the faculty to the signals and structure of incentives established by the State government, result in the need to take more drastic measures that would entail a permanent deterioration of the principle of academic freedom and teacher discretion in the implementation of the plans for classroom teaching. The next chapter explores the nature of critical racial theory and its impact on public education. Equally important is the role of the teacher in public education and his central position in the socialization of citizenship is explored in the last chapter of the volume. Although the American education system is diverse and decentralized, the case of Florida presents some of the central debates in American education.

Bibliography

The 1619 Project: Education Materials Collection. (2022). Retrieved from https://1619education.org/

Benlevy, S. (2000). Venus and Mars in the Jury Deliberation Room: Exploring the Differences that Exist among Male and Female Jurors during the Deliberation Process. *Southern California Review of Law and Women's Studies, 445*(Spring), 1-30.

Bloch, E. (2021). Duval County teacher Amy Donofrio terminated, settles lawsuit with school district. *The Florida Times-Union*. Retrieved from https://www.jacksonville.com/story/

news/education/2021/08/03/duval-schools-agrees-settlement-lawsuit-regarding-teacher-blm-flag/5477872001/
Chetty, R., Hendren, N., Jones, M. R., & Porter, S. R. (2018). Race and Economic Opportunity in the Unites States: An Intergenerational Perspective. *Race and Economic Opportunity in the United States: An Intergenerational Perspective, 135*(2), 711-783.
Civic Education Curriculum, Florida House of Representatives (2021 July 1). B5 Stat. Miami: Florida House of Representatives.
Cotton, J. (2009). Realism, Rationalism, Race: On the Early International Relations Discipline in Australia. *International Studies Quarterly, 53*(3), 627-647.
Hernandez-Armesto, F. (2003). *The Americas: The History of a Hemisphere.* London: Phoenix.
Florida Realtors Market Data. (2022). Retrieved from https://www. floridarealtors.org/newsroom/market-data
Gillion, D. Q. (2020). *The Loud Minority: Why Protests Matter in American Democracy.* Princeton: Princeton University Press.
Goldin, C. (1999). *A Brief History of Education in the United States.* Historical Paper 119 (August), 1-75. Retrieved from https://www.academia.edu/48959462/A_Brief_History_of_Education_in_the_United_States
Gonzalez, R. G. (2010). Ethnic Identity at a Majority Hispanic-Serving Institution. *Journal of Latinos and education, 9*(4), 284-302.
Hilburn, J., Journell, W., & Buchanan, L. B. (2016). A Content Analysis of Immigration in Traditional, New, and Non-Gateway State Standards for U.S. History and Civics. *The High School Journal, 99*(3), 234- 251.
Individual Freedom, The Florida Senate, HB 7 Stat. (2022 July 1, 2022). Miami: Florida Senate.
Jimenez, T. R. (2017). *The Other Side of Assimilation.* Santa Clara: University of California Press.

Justifiable Use of Force, 776.013 C.F.R. (2021).

Kamenetz, A. (2021). *Why education was a top voter priority this election*. NPR. Retrieved from https://www.npr.org/2021/11/04/1052101647/education-parents-election-virginia-republicans

Kaufman, E. (2019). The «1619 Project» Gets Schooled: The New York Times tries to rewrite U.S. history, but its falsehoods are exposed by surprising sources. *Wall Street Journal*. Retrieved from https://www.wsj.com/articles/the-1619-project-gets-schooled-11576540494

Kelin, E. (2020). *When Bipartisanship Becomes Irrational Why We're Polarized*. New York: Simon & Schuster.

Latest Statistics. (2022). Retrieved from https://floridajobs.org/work force-statistics/workforce-statistics-data-releases/latest-statistics

Lee, S. S., & Jimenez, J. (2021). Portrayal of Religion against the Back- drop of Progress and Modernity in the US and Canadian Social Science Textbooks from 1850 to 2021. In D. Hildebrandt-Wypych & A. W. Wiseman (Eds.). *Comparative Perspectives on School Textbooks*. Switzerland: Springer Nature, McKay Scholarship. Retrieved from https://www.fldoe.org/schools/school-choice/k-12-scholarship-programs/mckay/

Osorio, S. (2021). Senadores de EEUU presentan proyecto para pro- mover elecciones democráticas en Nicaragua. *El Nuevo Herald*, (March 28, 2021). Retrieved from https://www.elnuevoherald.com/noticias/mundo/america-latina/article250276030.html

Florida House of Representatives. (2022). *Parental Rights in Education*. HB 1557 Statute, July 1.

Pentón, M. J. (2021). Cuba construye una inmensa bandera de hormigón frente a la embajada de EEUU en La Habana. *El Nuevo Herald*, March 29. Retrieved from https://www.elnuevoherald.com/noticias/mundo/america-latina/ cuba-es/article250286590.html

Sabogal, E. (2012). Denaturalized identities: Class-based perceptions of self and others among Latin American immigrants in South Florida. *Latino studies, 10*(4), 546-565.
Shumow, M., & Pinto, J. (2014). Spanish-language immigrant media in Miami-Dade County, Florida: Discursive arenas for transnational civil societies. *The Latin Americanist, 58*(4), 59-83.
Tang, S. (2008). Fear in International Politics: Two Positions. *International Studies Review, 10*(3), 451-471.
Trischler, A. (2022). *Governor Ron DeSantis to approve $800 million to raise teacher salaries in Florida.* WTXL Tallahassee. Retrieved from https://www.wtxl.com/news/local-news/governor-ron-desantis-approves-800-million-to-raise-teacher-salaries-in-florida#:~:text=Monday%2C%20Governor%20Ron%20De Santis%20approved,to%209th%20in%20the%20nation
von Feigenblatt, Otto Federico. (2021a). *La Docencia en Tiempos de Pandemia*. Paper presented at the III Congreso Internacional Educación, Investigación y Tecnología «Innovación educativa y uso de las TI». Bogota, Colombia.
von Feigenblatt, Otto Federico (2021b). *Los Retos y Tendencias en la Educación Superior: Los Efectos la Pandemia de COVID-19.* Paper presented at the IV Congreso Internacional de Investigación e Innovación Universitaria y Simposio Estudiantil. Uniandes, extensión Quevedo, Ecuador.
von Feigenblatt, Otto Federico (2021c). Pensamientos sobre la Educacion. *Polimnia,* (23), 66-71.
von Feigenblatt, Otto Federico, & Acuña, B. P. (2021). Two Case Studies Dealing with Social Studies at the Secondary Level: Applied Anthropology and Grounded Theory. *Journal of Alternative Perspectives in the Social Sciences, 11*(2), 237-252.
von Feigenblatt, Otto Federico, Domíngue, D. C., & Vallé, J. E. G. (2015). The Family Educational Rights and Privacy Right (FERPA). *Journal of Alternative Perspectives in the Social Sciences, 6*(3), 329-335.

Whitten, S. (2022). *Disney vows to help repeal «Don't Say Gay» law, says Florida Gov DeSantis shouldn't have signed it*. CNBC. Retrieved fromhttps://www.cnbc.com/2022/03/28/disney-vows-to-help-repeal-dont-say-gay-law.html

2

Critical Racial Theory in Education: Controversies and Alternatives

There are a wide variety of critical theories, all originally derived from Marxism (Lemert, 2004; Willis, 2007). Traditional critical theory focuses on the unparalleled existence in societies of the impact that some institutions have in perpetuating social injustice (Stuart Sim, 2005; Williams, 1997). One of the characteristics of critical theory is that it rejects objectivity and firmly adopts the point of view of those it considers to be victimized by unjust structures. In other words, critical theory does not pretend to be impartial but rather to highlight injustices in support of marginalized groups (Minow, 1997). Thus, the many versions of critical theory differ from most theoretical frameworks traditionally related to research in the social sciences. Critical theorist combines social activism with research and thus integrates an important ideological aspect into the research process (Chodorow, 1997).

The first chapter of this volume introduced some central de- bates in American education, and in particular the controversy over legislative intervention in education. In particular, the debate about alleged police abuse in the

United States, and also about the role of the racial construct in education, has taken center stage in the country's social polarization. Racial Critical Theory is not new, it is one of the great branches of critical theory, along with critical gender theory and a wide variety of critical cultural theories (Bartky, 1997; Beauvoir, 1997; Charlesworth, 1997; Chodorow, 1997; Gottlieb, 1997; Holmstrom, 1997; Jaggar, 1997; Lemert, 2004; Minow, 1997; Murphy & Dingwall, 2007; Stuart Sim, 2005; Williams, 1997; Willis, 2007). Historically, critical racial theory has existed mostly in academic settings in universities, particularly in departments of so-called African American studies, something very peculiar to the United States but with little significance outside very small circles of intellectuals and political activists. Considered a radical theory in the social sciences with race as a construct, and rejected in anthropology and cultural sociology, it has regained new vigor in the United States due to its perceived usefulness for social activists (von Feigenblatt, 2015).

Before we can explain the nature of critical racial theory it is important to expose the history of the concept of race and its rejection by most social scientists. Partly a product of the classifying impulse of the European Enlightenment, the racial construct attempted to classify humans into racial groups. Humanity was initially divided into Negros, Mongoloids, and Caucasians. All three groups are very broad and include a wide variety of ethnic groups with physical and cultural characteristics that transcend continents and civilizations (von Feigenblatt, 2015). All three categories were further simplified outside academia in the American colonial dichotomy that classified people as black or white. The dichotomy described above derives in part from the system of slavery imposed on the British colonies of North America (Gearhart, 2005).

Historically, the definition of the «black» race has evolved until its heyday of elaboration during the era of segregation in the South of the United States at the beginning of the twentieth century (Chetty, Hendren, Jones & Porter, 2018). According to this classification system, having even a minimum percentage of black «blood» automatically means that the person is considered «black». In this case, paradoxically, phenotypic aspects are ignored, and the genealogy of the person is emphasized. A person may have pale complexion and be considered black if one of their grandparents was considered black (Gearhart, 2005). There are several explanations for the development of this system of racial classification. One of the most widely accepted focuses on the white minority's fear of losing control over a majority group (Hochschild, 1995). After the abolition of slavery, the white minority attempted to crystallize its dominance by maintaining a strict racial classification to limit access to public services and limit interaction within groups. The Jim Crow laws were based on the division of races, but only two existed from the legal and social point of view, particularly in the South of the United States (Hochschild, 1995; Hochschild & Powell, 2008).

The long and difficult struggle for civil rights initially focused on attacking the Jim Crow system and emphasizing the humanity of blacks to justify their inclusion in the social and political life of the country (von Feigenblatt, 2015). The goal of most civil rights groups was the erosion of the racial classification system. A central figure in the struggle for civil rights was Dr. Martin Luther King (von Feigenblatt, 2015). But it is also important to remember that other groups fought for an alternative view of the relationship between racial groups. Paradigmatic examples were Malcolm X and the terrorist group known as the Black Panthers. These groups accepted an alternative identity and internalized racial differences; a phenomenon known in Marxist theory as false consciousness, in this case from the point of view of racial identity. The difference is that the Black Panthers and to some extent also Malcolm X tried to forge an alternative identity to empower blacks but separated and in opposition to whites (von Feigenblatt, 2015). In the 60s of the twentieth century, Dr. Martin Luther King's version emerged as the victorious vision, in part because of its message of nonviolence and its promotion of the social and economic integration of blacks in the United States.

However, the alternate versions represented by Malcolm X and the Black Panthers have never totally disappeared from the public sphere. They endured marginally in groups of radical intellectuals and social activists. The issue of police violence and the supposed racial content of racial prejudice in the United States has given new impetus to socio-political perspectives normally considered marginal. It is difficult to explain why these marginal perspectives have resurfaced at the end of the second decade of the twenty-first century, but some contextual factors can be highlighted. Donald Trump's presidency was interpreted by some progressive groups as a direct attack on the consensus achieved during the decades after Martin Luther King's famous speech in Washington DC (Abramowitz, 2018).

President Trump included elements of racial nationalism in some of his speeches and received support from groups known as «white nationalists» (Abramowitz, 2018). Groups that emphasize the Anglo-European history of the United States as a central aspect of the country's development (Gillion, 2020). Trump's surprising popularity also scared liberal groups in urban centers traditionally controlled by the Democratic Party. Most of the media, traditionally controlled by liberal groups, also reacted to the frontal attack of Trump and his conservative allies (Gillion, 2020). An important example of this phenomenon is the case of Project 1619 presented in the first chapter of this volume (Harris, 2020; Kaufman, 2019). The New York Times, CNN, and other prominent outlets moved in a more ideological and distinctly liberal direction in an effort to counter the Trumpist movement, mostly focused on social media and Fox. The sheer resources of the largest media corporations propelled marginalized groups to the center of the national debate on issues of race and values. Isolated cases of police abuse received front-page coverage, and hours of live television were devoted to issues of social and racial justice (Harris, 2020). Thus, the radicalization of the civil rights struggle is partly a reaction to Trump's conservative populism.

For many radical academics focused on critical racial theory, it was a great opportunity to get out of the basement of the Ivory Tower and take advantage of the social and political situation to advance their careers. The case of Hannah-Jones is emblematic of this phenomenon (Harris, 2020). A New York Times journalist who spearheaded the 1619 project, reinterpreting U.S. history from an emphasis on the phenomenon of slavery, and who has a master's degree in journalism, and a minimal number of scholarly publications, but a large following due to the publicity received thanks to the support of the media and the New York Times. Hannah-Jones applied to receive a position as a professor at the University of North Carolina, initially her application was denied, but due to great pressure from the media and her thousands of followers, the University was virtually forced to give the position of professor with continuous contract to a person without a doctorate and with a minimum number of academic publications. Something impossible for anyone else with a lack of academic credentials. It is important to note that the University of North Carolina is a public institution that receives funds from taxes paid to the state of North Carolina. In other words, every North Carolina resident is indirectly paying Professor Hannah-Jones' high salary, a much higher salary than many other historians with doctorates.

Figures like Hannah-Jones receive constant media coverage on civil rights issues and in particular on aspects of education. Its visibility, combined with the perception of many parents that schools are being used to radicalize their children without ask- ing their permission, has made racial critical theory a highly sensitive issue in the United States. Many progressive activists and scholars present the inclusion of critical racial theory as part of the curriculum as an excellent way to transform American society and thus fight what they perceive as racism. The goal is to present critical racial theory as a new theoretical framework for reinterpreting not only American history but also civics in the country's schools and colleges.

Critical racial theory considers that race is a fundamental con- struct to understand the functioning of American institutions and that it is therefore important to reinterpret the socio-political and economic structure of the country from this point of view (Cotton, 2009; von Feigenblatt, 2015). According to the promoters of this theory, it is necessary to understand that racism is part of the legal and political system of the United States and that it is necessary to highlight these problems in order to promote meaningful reforms that can combat the racism that exists at all levels of American society. There are several levels of analysis in critical race theory, individual, group, and institutional. An important feature of some versions of critical racial theory is that individuals are recommended to confront their culpability in maintaining and supporting a system that incorporates racial and social injustice (Hochschild, 1995).

From the point of view of the integration of critical theory in the classroom, particularly in the primary and secondary classroom, this involves in many cases exercises and activities in which white students are instructed to face their guilt in the social injustices of the country. An implicit conclusion of critical racial theory is that dominant groups share the guilt of the system because they benefit from it. Due to the complexity of the analysis required to deconstruct some institutions at the national level through critical racial theory, in many cases conclusions are simplified for integration into the curriculum of primary and secondary schools. The result is that students are presented with a historical version full of guilt and oppression. Some activities tell white students that they are privileged and must accept their guilt. Although not all promoters of critical theory agree with the payment of reparations to the descendants of slaves, most do.

Although some of the ideas previously exposed have been dis- cussed in the university environment for a long time, and the introduction of ideas considered marginal in the academic dialogue is common, the great controversy is their integration into the school and collegiate curriculum because children do not have the tools to compare one theory with another. And they receive only the conclusions without understanding how those conclusions have been reached. One aspect that is particularly controversial and has caused great distress in some parents is that their children are forced at an early age to feel guilty for events that occurred long before their birth. The feeling of guilt is not pleasant, and therefore young children are suffering a feeling of guilt simply because their parents are «white» A hereditary classification system is being reintroduced. Children are automatically victims or victimizers. It is not yet known what effect this may have on children's metal development and in particular how it may affect the formation of a stable identity. Another important criticism is that a construct that was fought for a long time (von Feigenblatt, 2015) is being reproduced. If the racial construct is given centrality in social studies, the result will be that it will be internalized by students. It is a true irony that the descendants of those who fought to eliminate racial divisions today are struggling to reinforce them. Another aspect that also complicates the situation of guilt is that the descendants did not suffer slavery and the descendants of the plantation owners were not alive at the time the abuses were committed. Therefore, the new generations are being blamed for events in which they had no participation. They are implicitly being assigned a hereditary identity.

The possible payment of reparations would be punishing a generation for events that occurred centuries ago. Apart from that it is also being assumed that all African Americans today are the direct descendants of slaves and that all whites are the oppressors. Historical reality is much more complicated (Gearhart, 2005; Hochschild & Powell, 2008). Many African Americans today are descendants of immigrants from the Caribbean and many others have some white ancestry. In the case of the group classified as «white» we find an even more complicated situation due to the magnitude of immigration in the nineteenth and twentieth centuries.

Is it fair to blame the descendant of an Irishman who arrived in the late nineteenth century for colonial slavery? Or a descendant of a Jewish refugee who arrived from Germany in the 1930s? It is probably not the fairest thing to do. It is also reasonable to assume that the average teacher in a public school does not have access to their students' family trees and a guilt algorithm before presenting an activity incorporating critical racial theory into their classroom. The result at the individual level has been children who return home at the end of the day with a look of sadness and when parents ask them why, they answer that their teachers told them they are guilty of the country's racism.

Another problem with incorporating critical race theory into colleges and schools is the effect it can have on the development of a patriotic identity in the next generation. There is no doubt that the political system of the United States is not perfect, and that the founding fathers had weaknesses and defects just like the rest of humanity, but it is very dangerous to extrapolate the values of the present to evaluate historical figures. The founding fathers of the United States lived in a social and cultural context of their time (Fernandez-Armesto, 2003). They were intellectuals influenced by the academic and political currents of their historical context. Products of the Enlightenment, and part of epistemic communities that included some of the most important thinkers of the day, the founding fathers considered themselves innovative in relation to their context. The Constitution and the Declaration of Independence of the United States are documents that were considered progressive in their time and that have survived the test of time, telling countless generations of the deepest human aspirations for freedom and democracy.

It is also true that these documents excluded many social groups and that the system that was established did not have the same socio-political effect for all groups (Fernandez-Armesto, 2003). However, to conclude that the

American Revolution had as its central objective the protection of the institution of slavery, is not only an exaggeration but is historically incorrect (Harris, 2020; Kaufman, 2019). It is not necessary to teach children that the whole past is bad for them to learn to build a better future. It is very unfair to dismiss and reject all the sacrifices that past generations have made to highlight contemporary injustices, even if those injustices exist.

The issue of civic education in the United States is thus a highly contested field in the discipline of education. From the patriotism of World War II, to the skepticism of the Vietnam War period, and through the period of forced assimilation of the late nineteenth century, no consensus has been reached on this important part of primary and secondary education. In part due to the decentralized system of education in the United States, a topic that will be exposed in another chapter of this book, there is a great variety of models of patriotic and civic education in the country (Bronstein, 2020; Byrd & Varga, 2018; «Civic Education Curriculum», 2021). Among the most recognized models as archetypal of different philosophical currents can be mentioned the State of Texas and the States of New York. Although it is difficult to present the many differences between the two models, the main differences can be presented in broad strokes.

The State of Texas has a very peculiar history because it existed as an independent country before being voluntarily annexed to the United States (Fernandez-Armesto, 2003). Texas is considered a deeply conservative state and controlled mostly by the Republican Party. Texas schools teach U.S. history and a separate subject for Texas history. The anthems of Texas and the United States are also included as part of school activities. The history classes present a chronological and generally positive version of the socio-political development of the country with an emphasis on the sacrifices made by the armed forces in the different military conflicts. On the other hand, the goal of social studies in elementary and secondary education in the State of Texas is to instill in students a sense of patriotism and loyalty to the state and country. In the case of New York, critical racial theory has been integrated into the curriculum with the goal of giving voice to historically marginalized groups. In the case of New York, textbooks are included that are not approved in other states due to their critical and in some cases, sexual content. Most states are in between the two extremes previously presented.

The challenge of so many regional authorities to dictate the curriculum and implement the teaching of civic education in American public education has resulted in its politicization. The introduction of legislation to restrict the introduction of critical racial theory in classrooms is a reaction to the perception that many districts are using the theory as a theoretical framework for teaching social studies and also for the professional training of teachers in diversity issues. The first chapter of the book mentions an exemplary case in Florida.

It is important to understand that the critique of racial criticism is not its emphasis on diversity and inclusion, but rather its tendency to divide the population into ethnic groups and thus reify a construct that has been discarded by anthropology and that has caused great problems historically. Emphasizing the construct of race and reinforcing a student's identification with an ethnic group could have very serious effects on the country's socio-political future. While many countries are struggling with internal ethno-nationalisms after suffering bloody civil wars and in some cases, genocide, it is ironic that some progressive groups in the United States are promoting a theory that divides the population based on their race.

Another possible danger of integrating critical race theory into primary and secondary education is the potential psychological effects on students. More research is needed on aspects of identity development in relation to public blaming speech. In part young people are being asked to participate in rituals of public humiliation, and the most dangerous thing is that it is being done in the name of the alleged sins of their ancestors.

One of the biggest challenges in implementing the recommendations of racial critical theory at the school and college level is that students are not prepared to engage in conversations about sensitive topics. The spike in cases of psychological abuse in schools is troubling and adding such a comprehensive issue to a person's identity could result in a spike in the rate of teen suicide and violence. Facilitating conversations and dialogues on highly sensitive topics requires a lot of training and experience (Baker, 2005). It is important to remember that Racial Critical Theory emerged in universities and was introduced in advanced courses in the social sciences. Students in these courses not only had the maturity and practice to engage in dialogues on controversial issues but had also gone through a formative period in which their personalities and identities had been formed.

The case of primary and secondary school children, who are still in the process of formation and many of whom are suffering great challenges in the arduous process of developing their identities is very different. Talking about identity issues can be stressful especially for children who are in the process of exploring various facets of their emerging identities. Teachers, even though they are very well trained in teaching, may not have training in psychology to deal with situations that may arise in the process of talking about difficult topics. Much has been invested in student mental health in recent years, in part due to the increased violence in public schools and colleges. Many states' mental health resources are extremely strained and adding one more variable to students' mental health problems does not make much sense.

Recommendations

Critical Racial Theory is an interesting contribution to the academic dialogue on the effect of the construct called «race» and economic and political structures. As mentioned above, Critical Racial Theory is part of the traditional Marxist Theory of Critical Theory, but with a focus on «race». There is no doubt that Critical Theory has contributed to the development of the social sciences and that its focus on systemic injustice has contributed much to our knowledge about the socio-economic effects of certain economic regimes and systems (Chetty *et al.*, 2018; Davidheiser, 2008). Even so, it is important to remember that before criticizing a system or looking for weaknesses in it that system should be understood.

One recommendation, which also applies to other aspects of teaching, is further training of teachers in facilitation and moderation techniques for meetings. Critical Racial Theory can be included as a topic or theoretical framework of analysis in advanced social science courses in high school, but it must go together with other alternative interpretations of the same historical facts. The introduction of Critical Racial Theory as one among many other theories is a possibility. A strong emphasis on the ethical use of academic freedom must be integrated into teacher training so that teachers can separate their personal opinions and prejudices from their classroom teaching.

Another important aspect is that in primary education a strong foundation must be built from the point of view of civic education. Children should be taught how the system works and its objectives. The first step is for elementary students to understand the official version of patriotic civic education so that they learn about the positive side of the system. In this way, students would have the necessary preparation to understand not only the weaknesses of the system, but also its strengths. There is no point in inviting students to criticize a system they do not understand.

Bibliography

Abramowitz, A. I. (2018). *The Great Alignment: Race, Party Transformation, and the Rise of Donald Trump*. New Haven: Yale University Press.

Baker, L. L. (2005). Facilitator Core Competencies as Defined by the International Association of Facilitators. In S. Schuman (Ed.), *The IAF Handbook of Group Facilitation* (pp. 459-471). San Francisco: Jossey-Bass.

Bartky, S. L. (1997). From Feminity and Domination. In C. C. Gould (Ed.), *Gender: Key Concepts in Critical Theory* (pp. 103-112). New Jersey: Humanities Press.

Bronstein, E. A. (2020). «It isn't in the Curriculum»: World History Teachers' Views on U.S. Presidential Elections. *The Social Studies, 111*(3), 123-132.

Byrd, M., & Varga, B. A. (2018). The Manifestation of Campbell's Law: Consequences of Eliminating of Social Studies from the Curriculum. *Social Studies, 109*(1), 27-33.

Charlesworth, H. (1997). Human Rights as Men's Rights. In C. C. Gould (Ed.), *Gender: Key Concepts in Critical Theory* (pp. 384-393). New Jersey: Humanities Press.

Chetty, R., Hendren, N., Jones, M. R., & Porter, S. R. (2018). Race and Economic Opportunity in the Unites States: An Intergenerational Perspective. *Race and Economic Opportunity in the United States: An Intergenerational Perspective, 135*(2), 711-783.

Chodorow, N. J. (1997). Gender, Relation, and Difference in Psycho- analytic Perspective. In C. C. Gould (Ed.), *Gender: Key Concepts in Critical Theory* (pp. 25-40). New Jersey: Humanities Press.

Civic Education Curriculum, Florida House of Representatives, HB5 Stat. (2021 July 1).

Cotton, J. (2009). Realism, Rationalism, Race: On the Early International Relations Discipline in Australia. *International Studies Quarterly, 53*(3), 627-647.

Davidheiser, M. (2008). Race, Worldviews, and Conflict Mediation: Black and White Styles of Conflict Revisited. *Peace & Change, 33*(1), 60-89.

De Beauvoir, S. (1997). The Second Sex. In C. C. Gould (Ed.), *Gender: Key Concepts in Critical Theory* (pp. 3-15). New Jersey: Humanities Press.

Fernandez-Armesto, F. (2003). *The Americas: The History of a Hemisphere.* London: Phoenix.

Gearhart, R. (2005). Taking American Race Relations on the Road... to Africa. *African Studies Quarterly, 8*(2), 70-83.

Gillion, D. Q. (2020). *The Loud Minority: Why Protests Matter in Ameri can Democracy.* Princeton: Princeton University Press.

Gottlieb, R. S. (1997). Mothering and the Reproduction of Power. In C. C. Gould (Ed.), *Gender: Key Concepts in Critical Theory* (pp. 41-47). New Jersey: Humanities Press.

Harris, L. M. (2020). I helped Fact-Check the 1619 Project. The Times Ignored Me. *Politico.* Retrieved from Politico website:

https://www.politico.com/news/magazine/2020/03/06/1619-project-new-york-times-mistake-122248
Hochschild, J. L. (1995). Race, Class, and the Soul of the Nation: Facing up to the American Dream. Princeton: Princeton University Press.
Hochschild, J. L., & Powell, B. M. (2008). Racial Reorganization and the United States Census 1850-1930: Mulattoes, Half-Breeds, Mixed Parentage, Hindoos, and the Mexican Race. *Studies in American Political Development, 22*(Spring), 59-96.
Holmstrom, N. (1997). Do Women Have a Distinct Nature? In C. C. Gould (Ed.), *Gender: Key Concepts in Critical Theory* (pp. 56-65). New Jersey: Humanities Press.
Jaggar, A. M. (1997). Human Biology in Feminist Theory: Sexual Equality Reconsidered. In C. C. Gould (Ed.), *Gender: Key Concepts in Critical Theory* (pp. 48-55). New Jersey: Humanities Press.
Kaufman, E. (2019). The «1619 Project» Gets Schooled: The New York Times tries to rewrite U.S. history, but its falsehoods are exposed by surprising sources. *Wall Street Journal*. Retrieved from https://www.wsj.com/articles/the-1619-project-gets-schooled-11576540494
Lemert, C. (2004). *Social Theory: The Multicultural and Classic Readings*. Boulder: Westview Press.
Minow, M. (1997). Making all the Difference. In C. C. Gould (Ed.), *Gender: Key Concepts in Critical Theory* (pp. 397-407). New Jersey: Humanities Press.
Murphy, E., & Dingwall, R. (2007). The Ethics of Ethnography. In P. Atkinson, A. Coffey, S. Delamont, J. Lofland, & L. Lofland (Eds.), *Handbook of Ethnography* (pp. 339-351). London: Sage Publications.
Stuart Sim, B. V. L. (2005). *Introducing Critical Theory*. London: Icon Books.

von Feigenblatt, Otto Federico. (2015). The Fallacy of Race: A Post-racial America. *Journal of Alternative Perspectives in the Social Sciences, 7*(1), 39-53.
Williams, P. (1997). The Alchemy of Race and Rights. In C. C. Gould (Ed.), *Gender: Key Concepts in Critical Theory* (pp. 155-158). New Jersey: Humanities Press.
Willis, J. W. (2007). *Foundations of Qualitative Research: Interpretive and Critical Approaches*. London: SAGE Publications.

3

The American Decentralized Education System: Strengths and Weaknesses

Introduction

The first chapters of this volume have touched on some topics related to the importance of structure in educational leadership. Examples include the role of legislatures in developing content standards. The important role of school boards in implementing state laws at the local level has also been explained. This chapter presents a broader and more comprehensive exposition of the educational structure of the United States with an emphasis on its strengths and weaknesses. A theoretical framework focused on systems theories is applied to explore the complex interaction between different levels of government and between career bureaucrats and the many elected officials with authority on education issues.

One of the great challenges in the understanding of the complex functioning of the American educational system is the interaction between several open systems within it. State legislatures, various branches of the federal government, local school boards, professional teacher associations, and a wide variety of interest groups, *inter alia*, make up the education system in the United States. It should also be remembered that there are parallel systems of education at the K-12 parochial and private level again with specific characteristics.

Strengths

Much has been written about the weaknesses of the American education system, particularly in relation to low PISA test scores (von Feigenblatt, 2022a). Even so, it is important to recognize that the decentralized system also has great strengths and advantages over more centralized systems (Rangel, Ivanova, & Singer, 2009). To understand the strengths of the system it is necessary to present a brief discussion about its historical development and about the federal system of the United States of America. The process of development of the educational system cannot be separated from the process of political development of the country.

The United States was initially organized as a confederation of autonomous states and subsequently as a federation (Fernandez-Armesto, 2003; Wigman, 2021; Williamson, 1992). It should be emphasized that the initial thirteen colonies' voluntarily transferred part of their sovereignty and powers to the new federal government but retained many of their powers during the incorporation process. The constitution does not give the federal government authority over education (Zhao, Hoge, Choi, & Lee, 2007). According to the constitutional principle that powers not enumerated as assigned to the federal government belong to the states, education is considered a sector under the authority of individual states. The constitution does not mention public education as it did not exist, and universities emerged independently of state intervention (von Feigenblatt, 2021b). An excellent example is Harvard University, which was founded as a private, non-profit institution more than a century before U.S. independence (Goldin, 1999).

Public and private schools in the United States arose from localized efforts. There are three parallel systems of education at the K-12 level, the public education system (with the variant of semi-autonomous schools known as *charters*), the religious edu- cation system, and the secular private education system (Goldin, 1999). The three systems developed independently and have different structures. The religious system emerged organically be- fore independence, the public education system developed later, and finally the secular education system appeared.

Although the emphasis of this chapter is on the public education system, the structures, and origins of religious and secular education systems will be briefly explored. The religious education system is the oldest in the United States. Although we call it a system, it is made up of several independent systems. Several religious denominations established educational systems during the colonial period and then during the first years of independence (Zhao *et al.*, 2007). Although the model of the educational system varies from denomination to denomination, initially independent churches established affiliated schools. The schools operated privately funded and in many cases were subsidized by the church.

The largest religious education system in the United States is the Catholic college and university system. This system, according to US News and World Report, currently includes 5,938 institutions, a number that has decreased from a peak of almost 11,000 in 1970 (von Feigenblatt, 2022b). In higher education, there are currently 221 Catholic universities in the United States. Internationally famous institutions such as Notre Dame University, Boston College, Tulane University, and Georgetown University are part of the Catholic education system in the United States (von Feigenblatt, 2021a).

Other denominations with large educational systems include Baptists, Episcopalians, and Mormons. Because of the ease of establishing new religious denominations in the United States, there are many schools affiliated with small religious systems. There- fore, it is difficult to generalize from the structural point of view of the various educational systems (Chetty, Hendren, Jones, & Porter, 2018; von Feigenblatt, 2021b; Gonzalez, 2010; Hirst, 2005). Even so, some important characteristics can be identified. The mission and vision of these education systems focuses on religion (Gillion, 2020). On the other hand, being private institutions, school fees can vary from $9000 a year to $25,000. A major challenge with this type of system is that it relies directly on private funds from students' families. Property taxes and other taxes that partly help cover the costs of public education are not available and therefore families who send their children to religious education systems must pay their taxes in addition to paying the tuition of the private school (Mangu-Ward, 2008).

An important strength of religious education systems is that due to their financial autonomy, they have had to operate more efficiently than the public education system. A school that cannot cover its costs must close, which is not the case with public schools that may operate in deficit. On the other hand, in general, religious schools give great emphasis to the holistic education of the individual since their religious nature allows them greater freedom of teaching in matters of moral formation and values compared to secular and public schools (von Feigenblatt,

2020). In other words, traditional education incorporated aspects of instruction that went beyond the teaching of content, such as the teaching of appropriate behavior and values. Due to the great demographic diversity of the United States and the ideological polarization of the country, training in values and behavior has become a controversial issue in the public education system and in some secular schools (Baer, Wolf, & Risle, 1968; Hirst, 2005; Maag, 2008; Yoo, 2009).

Another strength of religious education systems is their selectivity. The level of selectivity of religious schools and colleges varies from institution to institution, but because they all have an economic cost and that enrolling a child has an extra process, the result is that families self-select and tend to include parents more committed to the education of their children than average. An indirect way that religious institutions use at the primary and secondary levels is to inform parents that they are expected to be active participants in their children's education, including supervising homework, communicating with teachers, and participating in school activities. The active inclusion of families in education gives them strength since education does not end in the classroom but continues at home.

The system of private schools and colleges in the United States is very varied and emerges from religious education systems (Thelin, 2011). A common process is that an institution begins its history affiliated with a religious denomination, but with the passage of time it loses religious emphasis. An example of this phenomenon is Harvard University which was established more than a century before the independence of the United States, with the original goal of training Protestant pastors, and gradually evolved into a secular university (Thelin, 2011). A similar process occurs in some schools and colleges in particular in cases in which institutions achieve economic independence from the churches that founded them. The process also depends on the initial relationship between the educational institution and the religious denomination. Schools and colleges directly run by a religious denomination and whose goal is to subsidize the education of a particular community tend to maintain their religious affiliation. In the case of elite schools and colleges that do not require any subsidy from the religious denomination, they can opt for autonomy and maintain a traditional orientation, but with an emphasis on secular education.

A more recent phenomenon is the proliferation of secular schools and colleges with an emphasis on alternative education methods. One sector has specialized in boarding schools for boys with behavioral and learning disabilities. Another sector is composed of Montessori schools and there is a wide range of schools and colleges with experiential learning methods (DeVos, 2022). The range and variety of institutional structures of secular private institutions make it difficult to generalize, but the conclusion that can be presented is that the secular private sector is characterized by its rapid adaptation to the market. It is a very volatile sector and most new institutions close their doors in less than a decade, yet the «creative destruction of capitalism» promotes innovation in education and adjusts to the needs of the market, since it depends on supply and demand. The adaptation and variety of educational options offered by the private sector in the United States is one of the best examples of the advantages of decentralizing the education system.

The public sector is divided between regular public schools and *charter* schools (DeVos, 2022; Goldin, 1999). The structure of the public education system is also highly decentralized. The ultimate authority over public education is each State's department of education. In turn, state governments delegate some of their authority to local school boards, divided by geographic districts. School boards appoint a superintendent who in turn serves as the executive branch of the school district and all schools and colleges in the district. Regular public schools are characterized by their standardization. All must follow the same curriculum, and teachers in such schools must meet state certification requirements. On the other hand, regular public schools tend to have a large investment in physical plant and also receive funding based on the number of students. There is a wide variety of public schools in the United States because there are approximately 13,000 school districts (DeVos, 2022). An advantage of this level of decentralization is that many districts adjust to the demographic and economic needs of the communities they serve. For example, schools in rural districts offer so-called «magnet» programs focused on agriculture while other districts offer a focus on auto mechanics. Therefore, it can be concluded that the structural decentralization of public education is key in adapting it to the needs of the local community.

Charter schools are public but enjoy great administrative autonomy (DeVos, 2022). A *charter* is basically an agreement or con- tract between a private group and the district. The process begins with a proposal to establish a semi-independent school or college. The group of people or organization applying for permission to open the *charter* school must provide the physical money and the proposal must explain the emphasis and type of curriculum that the school will follow. If approved, the charter school or college receives funding relative to the number of students but does not receive funding for physical plant. Unlike regular public schools, charter schools have more freedom to choose their curricula and can also choose how much to pay teachers and the types of incentives they offer. The logic of *charter* schools is to promote innovation in the public sector through administrative decentralization. The discretion and autonomy of *charter* schools is a great strength because it allows reforms that could not be made in regular public schools due to the bureaucracy of the system and also because of the collective agreements signed with the teachers' unions.

Weaknesses

The previous section focused on presenting some of the strengths of the U.S. decentralized system in the field of primary and secondary education. This section presents some weaknesses that are the counterpart of the strengths previously presented. In other words, the flexibility and decentralization of the system also includes a problematic side, lack of uniformity and coordination at the national level. The Department of Education at the Federal level has no direct authority over education at the national level and the same departments of education at the state level, although they have authority over education, in practice delegate their administration to school boards in the case of public education, and to the private sector in the case of religious and private education (von Feigenblatt, Acuña, & Cardoso-Pulido, 2022; Goldin, 1999; Kamenetz, 2021).

The result at the level of public policy is its fragmentation and in some cases its polarization. Fragmentation in the education sector is particularly problematic in cases of crisis at the national level. An example of this is the shortage of teachers at the beginning of the second decade of the twenty-first century. It is a problem at the national level that needs solutions at the national level. Unfortunately, not even the Federal Department of Education has managed to coordinate a public policy to deal with the problem at the national level. Secretary of Education Dr. Miguel Cardona has accepted the problem publicly and has simply recommended that school districts use resources provided by the Federal government during the COVID-19 crisis to seek local solutions to the severe teacher shortage (Czachor, 2022). The solutions presented by different districts and states have been very varied and poorly coordinated (*Identification of Critical Teacher Shortage Areas for 2022-23*, 2022). An example is the case of the State of New Mexico, in which the Governor mobilized the national guard to serve as substitute teachers in public schools and colleges (Lee & Attanasio, 2022). Another example is the Florida Governor's executive order to allow military veterans to teach with 60 college credits, the equivalent of half a bachelor's/degree (Jones, 2022). Although they are creative solutions, there is no state-to-state coordination to deal with a problem that does not recognize state or district boundaries.

An example of the problem of lack of coordination due to the decentralization of the education system in the United States is that an initiative in one district can cause a shortage in another district. In other words, due to the systemic nature of the teacher shortage, a seemingly successful initiative in one district can result in a short-term crisis in a neighboring school district. Many problems in the field of education require solutions at the macro level and the lack of coordination in the search for solutions can result in competition for a limited number of teachers, which leads to instability in human resources, without a sustainable solution. On the other hand, decentralization can also result in organizational oscillation. Lack of coordination between different school districts in a state, and between different states, results in organizational oscillation. In other words, structural changes in one district with the goal of initiating systemic reforms can be counteracted by organizational changes in other districts and vice versa.

Conclusions

This chapter has explored issues related to the decentralization of the U.S. education system, particularly with a focus on education at the elementary and secondary levels. The decentralization of the United States education system is a product of the country's historical development and reflects the great socio-political and cultural diversity of its population. The same flexibility that provides the system with the malleability needed to adapt to local needs is a major weakness when the system faces national and international crises, as in the case of teacher shortages. It is extremely difficult to coordinate public policies at the federal level because of the political structure of the United States, particularly in relation to education.

The decentralized nature of the system assumes that market pressure, combined with the influence of lobbyists, and the influence of constituents, will lead to effective and efficient public policies. Unfortunately, although the system is relatively flexible, the influence of teachers' unions and the bureaucracy of the school districts itself ossifies the administrative structure and hinders the implementation of public policies with popular support. The dilemma from the point of view of reform is between a truly decentralized and to some extent privatized system, and a more centralized system, but with better coordination. In other words, for decentralization to work, the market must be allowed to exert its influence free from distortions introduced by interest groups and directors' cartels. On the other hand, a centralized, French-style system needs sufficient authority and standardization to be effective.

Bibliography

Baer, D. M., Wolf, M. M., & Risle, T. R. (1968). Some Current Dimensions of Applied Behavior Analysis. *Journal of Applied Behavior Analysis, 1*(1), 91-97.

Chetty, R., Hendren, N., Jones, M. R., & Porter, S. R. (2018). Race and Economic Opportunity in the Unites States: An Intergenerational Perspective. *The Quarterly Journal of Economics, 135*(2), 711-783.

Czachor, E. M. (2022). Cardona calls on states to use COVID-19 relief funds to address teacher shortages. *CBS News*, August 21. Retrieved from https://www.cbsnews.com/news/miguel-cardona-covid-relief-funds-teacher-shortage-face-the-nation/

DeVos, B. (2022). *Hostages No More: The Fight for Education Freedom and the Future of the American Child*. Washington DC: Center Street.

Fernandez-Armesto, F. (2003). *The Americas: The History of a Hemisphere*. London: Phoenix.

Gillion, D. Q. (2020). *The Loud Minority: Why Protests Matter in American Democracy*. Princeton: Princeton University Press.

Goldin, C. (1999). A Brief History of Education in the United States. *Historical Paper 119*, 1-75. Retrieved from https://www.academia.edu/48959462/A_Brief_History_of_Education_in_the_United_States

Gonzalez, R. G. (2010). Ethnic Identity at a Majority Hispanic-Serving Institution. *Journal of Latinos and education, 9*(4), 284-302.

Hirst, R. K. (2005). Reducing Discipline Referrals in Middle School.

Principal, 84(3), 51-51.

Identification of Critical Teacher Shortage Areas for 2022-23. (2022). Tallahasee: Florida Department of Education. Retrieved from https:// www.fldoe.org/core/fileparse.php/20042/urlt/7-2.pdf

Jones, J. H. (2022). Why Florida's plan to recruit veterans as teachers is troubling. *MSNBC,* (August 16, 2022). Retrieved from https://www. msnbc.com/the-reidout/reidout-blog/florida-veterans-teachers- desantis-rcna43311

Kamenetz, A. (2021). Why education was a top voter priority this election. *NPR*. Retrieved from https://www.npr.org/2021/11/04/10521 01647/education-parents-election-virginia-republicans

Lee, M., & Attanasio, C. (2022). New Mexico asks Guard to sub for sick teachers amid Omicron. *ABC News,* January 19. Retrieved from https://abcnews.go.com/Health/wireStory/mexico-asks-guard- sick-teachers-amid-omicron-82362439

Maag, J. W. (2008). Rational-Emotive Therapy to Help Teacher Control Their Emotions and Behavior When Dealing with Disagreeable Students. *Intervention in School & Clinic, 44*(1), 52-57.

Mangu-Ward, K. (2008). Education for Profit. *Reason, 40*(3), 38-45.

Rangel, E., Ivanova, A., & Singer, B. (2009). Higher education, employment and globalization: The cases of Mexico and Thailand. *Pacific-Asian Education Journal*, 75-90.

Thelin, J. R. (2011). *A History of American Higher Education* (2nd ed.). Baltimore: The John Hopkins University Press.

von Feigenblatt, Otto Federico. (2020). The Importance of Historical Heritage and the Fallacy of the Cancel Movement: International Case Studies. *Journal of Alternative Perspectives in the Social Sciences, 10*(3), 483-492.

von Feigenblatt, Otto Federico (2021a). *Los Retos y Tendencias en la Edu- cación Superior: Los Efectos la Pandemia de COVID-19.* Paper presented at the IV Congreso Internacional De Investigación E Innovación Universitaria Y Simposio Estudiantil. Uniandes, Extensión Quevedo. Ecuador.

von Feigenblatt, Otto Federico. (2021b). Pensamientos sobre la Educacion. *Polimnia* (23), 66-71.

von Feigenblatt, Otto Federico. (2022a). La enseñanza de la Historia de los Estados Unidos en el Estado de la Florida: Una interpretación de la teoría fundamentada. *Dilemas contemporáneos: Educación, Política y Valores* (1), 1-27. Retrieved from file:///C:/Users/vonfe/Downloads/3238- Texto%20del%20art%C3%ADculo-4432-1-10-20220811%20(1).pdf

von Feigenblatt, Otto Federico. (2022b). La enseñanza de la Historia de los Estados Unidos en el Estado de la Florida: Una interpretación de la teoría fundamentada. *Revista Dilemas Contemporaneos X* (1), 1-25. Retrieved from https://dilemascontemporaneoseducacionpoliticayvalores.com/index.php/dilemas/article/view/3238

von Feigenblatt, Otto Federico, Acuña, B. P., & Cardoso-Pulido, M. J. (2022). *Nuevos paradigmas didácticos y otras proximaciones: aprendiza- je personalizado y Education Maker.* Barcelona: Octaedro.

Wigman, B. W. (2021). *A Concise History of the Caribbean* (2nd ed.). New York: Cambridge University Press.

Williamson, E. (1992). *The Penguin History of Latin America.* London: Penguin Books.

Yoo, J.-J. (2009). Peer Influence on Adolescent Boys' Appearance Management Behaviors. *Adolescence, 44*(Winter), 1017-1031.

Zhao, Y., Hoge, J. D., Choi, J., & Lee, S.-Y. (2007). Comparison of Social Studies Education in the United States, China, and South Korea. *International Journal of Education, 21*(2), 91-122.

4

Higher Education for Hispanics in the United States

Introduction

The Hispanic community in the United States has grown considerably during the last few decades (Frey, 2020). But its importance transcends numerical indicators due to the impact that many of its members have had on different sectors of society (Chetty, Hendren, Jones, & Porter, 2018; Gonzalez, 2010). Currently, the Hispanic community represents approximately 18 percent of the population of the United States (Frey, 2020). Thus, Hispanics represent the largest minority in the Country, followed by blacks at 12.2% of the population, and Asians at 5.6 percent (Frey, 2020). Another important characteristic of the Hispanic community is its youth. In other words, compared to other demographics, most Hispanics tend to be young, which has a direct impact on higher education.

This chapter explores some important issues related to higher education for Hispanics in the United States. Starting with demographic changes that project that Hispanics will continue to grow as a percentage of the population in the United States and exploring the issue of vocational education and formal education, and finally discussing the role of Spanish in higher education. At the end of the chapter, some challenges and opportunities identified with corresponding tentative recommendations are presented.

Demographic Changes

The Hispanic population in the United States has become the largest demographic minority (Frey, 2020). Not only has it dis- placed the African-American community as the largest minority community, but growth projections indicate that it will continue to grow at a faster rate than other minorities (Chetty *et al.*, 2018). It is also important to consider that the increase is not only due to immigration, but more than anything depends on the growth rate of the domestic population of Hispanics (Saldaña, 2021). Another important factor is the origin of immigrants. Historically, much emphasis has been given to immigration from Mexico, but today immigration from other parts of Latin America has increased, particularly immigration for political reasons (von Feigenblatt, 2022). It should also be mentioned that immigration includes a wide variety of social classes and demographic groups, which results in a great diversity of Hispanics in the country (Sabogal, 2012).

In the field of education, a slight change in the demographic composition of Hispanics can have important consequences, particularly in higher education (Gonzalez, 2010; Saldaña, 2021). A relatively recent trend is the immigration of highly qualified people for political reasons. Although this type of immigration is not new, it has increased in part due to the political crisis in Venezuela and due to the economic crisis in Latin America caused by

COVID-19. Although the percentage of immigrants with a college degree is low compared to most immigrants, this percentage has a major impact on American higher education. According to the Pew Research Center, there are large differences between states in relation to the education of Hispanic immigrants (Mora, 2022). For example, in Michigan 43% of immigrants have at least a bachelor's degree while in California 19% have a bachelor's degree. It is important to note that 35% of adults over the age of 25 in the United States have a bachelor's degree and therefore the educational level of immigrants is similar to the educational level nationally in the United States (Noe-Bustamante, 2020).

Another interesting observation is that the percentage of Hispanics with bachelor's degrees is 20.6% while the percentage of African Americans with bachelor's degrees is 28.1%, that of Anglos is 41.9%, and that of Asian Americans is 61.0% (Noe-Bustamante, 2020). Considering that the Hispanic minority is the largest minority and the one that is growing the fastest for reasons of immigration and domestic repro- duction, the Hispanic population has great potential from the point of view of higher education. Assuming that the Hispanic minority is in the process of reaching the national average of degrees at the undergraduate level, 35% of the population, means that, in the coming years, *ceteris paribus*, will see an increase from 20 to 35%, which means that the percentage of Hispanics with college degrees can increase by at least 15%, a higher proportion than any other demographic.

Taking into account the demographic trends previously ex- posed, some tentative interpretations of implications for higher education in the United States can be reached. First, it is clear that there are a large number of immigrants with university degrees obtained in their countries of origin. Another interpretation is that the average per capita income of the Hispanic community in the United States is not as high as that of other ethnic groups, there is the potential for economic growth for this important minority. There is an inverse relationship between undergraduate studies and economics. It's counterintuitive, but when the economy is booming, applications for college admission, particularly at the undergraduate level, go down, while when the economy is in recession, applications increase. This can be explained because during times of economic recession unemployment usually increases and many people without undergraduate studies invest in university studies to improve their chances of getting a job. On the other hand, in times of low unemployment and a fast-paced economy, many people can find good jobs without needing a bachelor's degree and therefore the number of applications goes down.

Due to the demographic trends previously exposed, we can tentatively conclude that the demographic known as Hispanic will become increasingly important to universities at the under- graduate and graduate level. One reason is that the number of young Hispanics will continue to increase due to domestic re- production and immigration. Considering that the growth of other demographic groups is slow, that unemployment is currently low in the United States, it is very likely that the proportion of Hispanic students in degree programs will increase considerably (Mora, 2022). Thus, the needs and characteristics of this important demographic group will take center stage in the strategic planning of the higher education sector in the United States.

Another interesting phenomenon is the increase in the number of foreign professionals who are immigrating to the United States with undergraduate degrees (Mora,

2022). There is an interesting social stratification among professionals who are immigrating from Latin America to the United States. Elites tend to speak English, but the middle class and lower middle class do not speak English well, although they do have undergraduate degrees from their home countries (Sabogal, 2012). This phenomenon combined with an increase in the number of businesses focused on the Hispanic community has resulted in the creation of jobs with little need to speak in English and that are almost entirely Spanish. These types of jobs exist in enclave regions with Hispanic majorities. In this type of enclave region almost all businesses are focused on the Hispanic community and the main business language is Spanish.

While it is nothing new that communities with Hispanic majorities exist, an important change is that many of these communities include highly skilled jobs due to the investment of successful Hispanics at the domestic level and the investment of members of the newly arrived elites. A paradigmatic example of this type of ethnic enclave is the City of Doral in South Florida (Shumow & Pinto, 2014). This city has a high growth rate and a Hispanic majority. A special feature of the city is that real estate is expensive, with large developments of luxury suburban communities, the concentration of Hispanic businesses in industries that require highly skilled labor, and the use of the Spanish language as the primary language in the city. One result of the development of this type of Hispanic enclave is that many professionals can practice without the need to speak much English. Due to the critical majority of Hispanics in these regions and that employers also speak Spanish, the use of the language is maintained from generation to generation and many of the professionals who are in this type of employment situation prefer to pursue postgraduate studies in Spanish instead of having to improve their English first to then be able to take postgraduate courses in the United States (Shumow & Pinto, 2014).

Gradual Transition from Vocational Studies to Formal Education

Due to the demographic trends presented in the previous section, a trend in higher education is the transition from vocational studies to formal education. Although many Hispanic immigrants are unskilled labor and many do not have basic education in their countries of origin, a large number of recent immigrants of middle class and lower middle class origin have some undergraduate studies and bring with them the sociocultural expectations of their class (Sabogal, 2012). In other words, for many re- cent immigrants the American dream is much more than just making money and hoping to pursue professions aligned with their class identity. Therefore, many immigrants seek formal edu- cation in the United States and not just vocational studies. One of the reasons that explains this transition is the prejudices that exist in many Hispanic countries against vocational education and in favour of formal education (Fernandez-Armesto, 2003).

The literature in the field of social stratification clearly identifies the constancy of this phenomenon in middle- and lower-middle-class immigrants (Pieterse, 2007). On the other hand, the domestic Hispanic population, those born already in the United States, in some cases have achieved economic success as small contractors and many also have expectations of achieving professional success and not only economic. The result has been a slow but sure transition from Hispanic education to formal education. There are a wide variety of sociocultural factors for the transition previously exposed. Hispanic communities are interconnected with countries of origin and with networks of recent im- migrants (Williamson, 1992).

The phenomenon of enclaves dominated by Hispanic companies reproduces many sociocultural structures of the countries of origin of the companies and the professionals who lead them. Hispanic-focused media tend to belong to Hispanic multinationals based in Mexico and other countries in the region (Shumow & Pinto, 2014). Therefore, and partly due to the acceleration of the process of globalization and communication technologies, there are transnational spaces with cultural characteristics very different from those of other ethnic groups such as Anglo-Americans. A strong economy combined with a constant cultural pressure to obtain university degrees for the prestige that comes with formal education offers an important opportunity for the American higher education sector.

Given the above, a significant increase in Hispanic students in undergraduate liberal arts programs and formal education at the graduate level can be expected. This is not to say that vocational programs will lose their importance, as for a majority of newly arrived and working-class Hispanics, high salaries in technical careers combined with the short time it takes to complete a technical program will continue to be very attractive. The main change will be the gradual increase in formal programs focused on Hispanic education. This is something that has already been recognized at the federal level with the denomination of «Hispanic serving institution» for universities (Gonzalez, 2010). Another change will be the proliferation of graduate programs in Spanish for Hispanics residing in the United States and also to attract Hispanics in the rest of Latin America.

The Role of Spanish in Higher Education

The role of Spanish in American in higher education is one of the most important unknowns of our times (Shumow & Pinto, 2014). Hispanic identity focuses on the use of the Spanish language and therefore from the anthropological point of view one of the most important unknowns is whether the use of Spanish will continue to increase in the United States or if it will be slowly lost due to the assimilation process. We must consider that the term «Hispanic» focuses on the use of the language and therefore the loss of the Spanish language could result in a dilution of Hispanic identity.

Although it is very difficult to predict the future of the Span- ish language in the United States, we can be guided by some cur- rent trends and by the experiences of other countries such as Canada. There are examples of minority communities that have managed to maintain their language in countries with majorities that speak another language. The best known case is that of the Quebecois in Canada, an ethno-cultural group that has managed to maintain the use of French in a country where English is mostly spoken (Abulof, 2009).

From the point of view of higher education, the use of the Spanish language is an important aspect considering that American higher education is said to be the best in the world and therefore offering official bachelor's, master's, and doctoral programs increases international competition for students with the Spanish language as a first language («World University Rankings», 2022). In other words, in a highly competitive market for international talent, if American universities offer degrees in Spanish, this puts them in direct competition with institutions in Spanish-speaking countries that also offer those degrees.

Another important effect of the possible increase in the use of Spanish in American higher education is the increase in research published in Spanish. Because American universities enjoy central positions in epistemic networks, publishing in Spanish by scholars based on American universities would give them virtu- ally immediate centrality. A parallel example from which we can extrapolate some lessons is the music industry. Many Hispanic artists who become internationally recognized managed to make the international leap while based in the United States due to access to influential distribution networks. Something very simi lar could happen, and is already happening in isolated cases, with Hispanic scholars based in the United States. The main change would not be the centrality of Hispanic scholars based in the United States, the case of Manuel Castells is a great example of the structural advantages of being affiliated with an American university, but rather the serious change from research published in English to publications published in English and Spanish.

From a research standpoint, another possible predictable effect will be closer collaboration between Spanish-American universities and U.S. universities. The collaboration between Latin American educational institutions and universities in the United States is already quite extensive, but the increase in research published in Spanish may give greater centrality to this relationship (Meenakumari & Krishnaveni, 2011). A reason why it is considered that this relationship will take on greater centrality in the coming decades is the large number of statesmen and business leaders in Latin America who have studied in the

United States. Another reason is the importance of the US economy to the Americas, with the prolonged recession and energy crisis in Europe, it is likely that we will see a contraction in investment in higher education and in particular research and therefore the relationship with higher education in the United States will be- come even more important. Therefore, it is very likely that the increase in the use of Spanish in research in American higher education will result in a growing centrality of U.S. research centers in Latin America, a leadership that until now has been limited in part due to language difference.

Challenges and Opportunities

Higher education for Hispanics in the United States is in a process of profound change. The changes depend on internal and external factors since the relationship of movement of people and ideas between Latin America and the United States has intensified in recent years due to political changes in the region. One of the big unknowns is the effect that the increase in the percentage of Hispanics in the United States can have from the point of view of the use of the language for education and research. A critical mass of U.S.-educated Spanish speakers can have a transformative effect on American higher education in relation to programmatic offerings and from the point of view of the subject matter and emphasis of the research.

Although possible trends have been presented in this chapter with the premise of *ceteris paribus*, possible changes in sociopolitical trends in Latin America and the United States could change the expected trajectory of development of higher education. For example, an increase in economic immigration to the United States by low-skilled

Hispanics could shift the emphasis to vocational training rather than formal education. Another important factor is the permanence of Spanish in the Hispanic community in the United States. Language permanence depends on a wide variety of factors, economic opportunities, socio-cultural aspects, and public policies. Another important factor for the permanence of Spanish is the speed of reach of a critical mass of His- panics in the United States. The speed of increase of the Hispanic population is a factor of great importance since a gradual increase can result in the dilution of the language and in the possible complete assimilation of the Hispanic community. On the other hand, a rapid increase can help reach a critical mass of Spanish speakers and thus perpetuate the permanence of Spanish in American higher education.

From the point of view of challenges, one of the most obvious challenges is the shortage of qualified teachers who can teach in Spanish. The Hispanic community is growing rapidly but the faculty does not reflect that diversity. Approximately 10 percent of the faculty is Hispanic, but not all have an appropriate level of Spanish to teach at the university level. Although the solution is apparently simple, allowing professors from Latin America to teach at American universities is not a feasible solution due to differences in the quality of education at the doctoral level. It is an open secret that American universities prefer professors with terminal degrees obtained in American universities. Obviously, the situation is different for professors with terminal degrees from elite foreign universities. The challenge is that Latin American universities have not achieved great success in international rankings, which hinders their integration into the faculty of American universities at the national level and therefore the increase in the number of courses and programs offered in Span- ish has resulted in a severe shortage of professors («World University Rankings», 2022). Although many Hispanic students are pursuing degrees in the United States, but very few continue to the doctoral level required to be a university professor, and of those not all opt for academic careers.

Tentative Conclusions

This chapter has explored several issues of importance in Ameri- can higher education from the Hispanic point of view. The main conclusion is that the increase in the Hispanic population in the United States will have a major effect on the development of higher education in the country. Another important conclusion is that support teams and trained teachers will be needed to deal with this important socio-cultural group. The changes will, *inter alia*, include changes to curriculum design, the development of new programs, and changes in institutional relations.

There is no doubt that more research is needed on the possible effects the increase may have on Hispanic students. The possible effect of the use of Spanish in higher education programs in the United States is also of great interest, not only from the point of view of curriculum development, but also from the point of view of epistemic networks and knowledge development. Spanish is central to Hispanic identity and therefore its permanence is the central factor that will determine many of the trends presented in this chapter.

Bibliography

Abulof, U. (2009). «Small Peoples»: The Existential Uncertainty of Ethno-national Communities. *International Studies Quarterly, 53*(1), 227-248.

Chetty, R., Hendren, N., Jones, M. R., & Porter, S. R. (2018). Race and Economic Opportunity in the Unites States: An Intergenerational Perspective. *The Quarterly Journal of Economics, 135*(2), 711-783.

Fernandez-Armesto, F. (2003). *The Americas: The History of a Hemisphere.* London: Phoenix.

Frey, W. H. (2020). The nation is diversifying even faster than predict- ed, according to new census data. *The Brookings Institution.* Retrieved from https://www.brookings.edu/research/new-census-data-shows-the-nation-is-diversifying-even-%20faster-than-predicted/

Gonzalez, R. G. (2010). Ethnic Identity at a Majority Hispanic-Serving Institution. *Journal of Latinos and education, 9*(4), 284-302.

Meenakumari, J., & Krishnaveni, R. (2011). Transforming Higher Educational Institution Administration through ICT. *International Journal of Advanced Computer Science and Applications, 2*(8), 5-54.

Mora, L. (2022). *Hispanic enrollment reaches new high at four-year colleges in the U.S., but affordability remains an obstacle.* Retrieved from https://www.pewresearch.org/fact-tank/2022/10/07/hispanic-enrollment-reaches-new-high-at-four-year-colleges-in-the-u-s-but-affordability-remains-an-obstacle/

Noe-Bustamante, L. (2020). *Education levels of recent Latino immigrants in the U.S. reaches new highs as of 2018.* Retrieved from https://www.pewresearch.org/fact-tank/2020/04/07/education-levels-of-recent-latino-immigrants-in-the-u-s-reached-new-highs-as-of-2018/

Pieterse, J. N. (2007). *Ethnicities and Global Multiculture: Pants for an Octopus* (Hardcover ed.). Plymouth: Rowman & LIttlefield Publishers, Inc.

Sabogal, E. (2012). Denaturalized identities: Class-based perceptions of self and others among Latin American immigrants in South Florida. *Latino studies, 10*(4), 546-565.

Saldaña, L. P. (2021). The Struggle for Mexican American Studies in Texas K-12 Public Schools: A Movement for Epistemic Justice through Creation/Resistance. *Association of Mexican American Educators (AMAE) Journal, 15*(2), 9-35.

Shumow, M., & Pinto, J. (2014). Spanish-language immigrant media in Miami-Dade County, Florida: Discursive arenas for transnational civil societies. *The Latin Americanist, 58*(4), 59-83.

von Feigenblatt, Otto Federico (2022). La enseñanza de la Historia de los Estados Unidos en el Estado de la Florida: Una interpretación de la teoría fundamentada. *Revista Dilemas Contemporaneos X*(1), 1-25. Retrieved from https://dilemascontemporaneoseducacionpoliticayv alores.com/index.php/dilemas/article/view/3238

Williamson, E. (1992). *The Penguin History of Latin America*. London: Penguin Books.

World University Rankings. (2022). Retrieved from https://www.times highereducation.com/world-university-rankings

5

Structural and Contextual Factors Exacerbating the Teacher Shortage in Florida

Introduction

The United States is facing a severe teacher shortage (Czachor, 2022). Numbers vary from state to state but Florida, an affluent southern state is representative of the challenges faced by many regions in the United States. There is considerable discussion over the possible causes of the teacher shortage. Common arguments include, *inter alia*, insecurity in schools, low teacher pay, low teacher morale, and burdensome standardized testing requirements. Nevertheless, a more nuanced understanding is needed to foster informed public policy. No single factor can adequately explain the severe teacher shortage in Florida and thus policies dealing with a single factor will invariably fail.

The present exploratory study attempts to bridge the lessons learned from different disciplines and perspectives which have tackled the thorny problem of teacher turnover and recruitment. At the micro-level the focus has been to understand the lived experiences of teachers and in particular, their levels of satisfaction or dissatisfaction with their careers (Anderson, 2006; Kantabutra & Saratun, 2013). Nevertheless, it is dangerous to extrapolate lessons from personal experiences, for a macro level phenomenon such as the nationwide teacher shortage. Therefore, a good starting point is to explore structural elements and how they may affect the incentive and disincentive structure to enter and to continue in the teaching career.

A focus on the decentralized structure of the American education structure (Stern & Johnston, 2013; Suttichujit, 2009; Zhao, Hoge, Choi, & Lee, 2007). A closer look at the complex interaction between elected officials at the Federal, State, and district levels provides a partial explanation for the lack of coordination to tackle the problem. Moreover, structure interacts with contextual factors such as demographic changes as well as the economic and political environment. A section of the paper explains some of the main opportunities and threats presented by the socio-economic and political environments which have exacerbated the crisis. Finally, specific case studies of strategic blunders are explored in relation to education policy in the state of Florida. Emphasis is placed on policies affecting teacher recruitment and retention.

Structural Factors

The American education system is highly decentralized (Goldin, 1999). Local elected school boards interact with state legislatures and governors, who in turn must deal with Federal level public policies. It should be noted that the United States developed during the height of the enlightenment period; the founding fathers had an aversion to centralized control over culture and religion. As a result of the ethos of the time and of the unique early historical development of the United States, the Constitution enshrined several important principles that would lead to a highly decentralized education system. One of the central principles is that the powers not explicitly reserved for the Federal Government belong to the State governments (Anderson, 2006). In the case of education, it is not explicitly mentioned as one of the powers of the Federal Government, and thus it is widely accept- ed that the individual state governments have authority over the regulation of education at both the K-12 and tertiary levels. Moreover, state legislatures have the power to determine the standards taught in schools (Hird, 2004).

Another layer of administration and structure is the local school boards (Saldaña, 2021). As previously stated, the Ameri- can education system, at the k-12 level, is highly decentralized. School boards have control over school districts, the basic ad- ministrative unit in the American education system. Members of the school boards are elected by local constituents and serve four-year terms. The role of school boards is akin to a legislature for the school districts. School boards have power over budgets, appointment of superintendents, principals, and teachers, as well as power over local guidelines for schools. Thus, local school boards occupy a central place in the administrative structure of the school system. Superintendents are usually appointed and answer to elected school boards and serve as the executive branch of the school district. It should be noted that Florida is one of only two states that allow some school districts to have elected superintendents. Nevertheless, the trend has been to move away from elected superintendents to appointed ones.

Therefore, the structure of the local school district reflects the political structure at the national and state levels with three branches of government, namely the executive, legislative, and judicial. In the case of school districts, the judicial function is exercised by local district courts and by county courts depending on the case. School boards have legal officers to handle and ad- vice the board regarding legal issues. The school board serves as a legislative branch and the superintendent, and his staff serve as the executive branch. In most districts elections to the school board are done by sub-districts. In other words, a school district is geographically divided into several geographic regions and each region elects one member to the school board. Technically school board members are non-partisan, however in practice candidates have the informal support of one party or another. There are no minimum educational requirements to run for the school board. Any person who is 18 or older and who does not have a criminal record can run to be a member of the school board. This is important, because it means that in contrast to other local elected positions such as for county level judge, many school board members have no experience in education, nor do they necessarily hold advanced degrees in the discipline.

There is also an important structural factor which affects the incentive structure to run for office. Serving in a school board is not considered to be a full-time job and therefore the remuneration for the service is relatively low. Salaries for schoolboard members range from $40,000 to $60,000 depending on the district. The current governor of Florida, Ron DeSantis, has even proposed making the position completely pro bono and thus remove the financial incentive to run for office. Currently it is possible to hold a full-time job or to own a private business while serving in the school board. Nevertheless, it is not possible to serve in the school board and at the same time work for the school district in another capacity. The result of this rule is obvious, teachers and principals cannot be members of the school board without having to leave their jobs.

Most collective bargaining agreements include special accommodations for teachers who want to run for political office. Nevertheless, even if the teacher or principal can request a leave of absence, the salary issue remains an important disincentive to run. If a teacher decides to run for office, he or she can take an extended leave of absence and his or her job will be protected, nevertheless the person will not receive his or her usual salary while serving on the school board. The newly elected school board member will be receiving the $40,000 to $50,000 school board member salary. Taking into consideration that the starting salary for a teacher in Florida is $47,500 and that due to the high cost of rents in some part of the state, it is difficult to recruit teachers for open positions, because of the low pay, it is unrealistic to expect that a teacher would give up his or her pay in exchange for serving in the school board and thus receive a pay cut rather than a raise.

On the other hand, a small business owner or a professional in a different field can keep his or her income and add the extra $40,000 to $50,000 to their yearly income. Thus, there is a disincentive for people with experience and knowledge about education to run for seats in the schoolboard. This results in the anomalous situation in which most school board members lack experience and academic credentials in the field while teachers are not as politically active as in other countries, for example the countries of the European Union (Bronstein, 2020).

Another structural factor that should be discussed is the appointment of the superintendent and other administrative staff. In the State of Florida there are a few state-mandated rules for the appointment of such personnel. For public schools, administrative officers such as principals, assistant principals and super- intendents are required to hold a master's or doctoral degree in educational leadership from a regionally accredited university and to pass the Florida Educational Leadership Examination (FELE). Three years of teaching experience are required as a minimum and there is no requirement for the teaching experience to be at the level of education the person intends to lead.

Thus, it is possible for a former lower schoolteacher to become a principal at a high school. It is also possible to become a principal or assistant principal without experience teaching a core subject. One such example is an assistant principal who taught home economics for three years, a subject that has been phased out since that time and is now in charge of conducting teacher evaluations at a middle school. Another structural challenge is that some school districts limit the hiring of administrative personnel to people already in the school district. Thus, the positions are not subject to an open search but rather appointment is done internally through a complex and obscure system involving public consultation with the community but ultimately decided by area-superintendents. Therefore, the most common incentive structure for advancement in Florida school districts discourages many ambitious teachers from continuing in the teaching career track and instead favor entering the administrative track.

Further explanation about the twin track in public education is necessary. A new teacher has two possible career tracks to consider in most public-school districts in the State of Florida. One career track focuses on teaching and leads to the rank of department head or lead teacher. Salary raises for teachers in the teaching track are based on years of service and formal credentials, supplements for advanced degrees. On the other hand, a teacher may choose to enter the administrative track after just three years of teaching. To do so, the teacher needs to pursue a master's degree in educational leadership at a regionally accredited university and then pass the Florida Educational Leadership Examination (FELE). Those are the official minimum requirements, nevertheless the incentive structure for entry into the administrative track favors teachers who prefer non-teaching duties.

While there are many pathways of entering the administrative track beyond the minimum requirements, the usual path begins with a teacher expressing an interest to an assistant principal or principal about their career goals. Over the years, districts have added a range of non-teaching duties to teachers, such as patrolling the hallways, overseeing security during lunchtime, and helping organize standardized testing. Some schools even require teachers to patrol the main door of the school before and after school. The first three years of teaching are the proving ground for many young graduates (Stern & Johnston, 2013). Many young teachers realize that they do not like teaching or sadly they cannot adapt to the reality of classrooms in many public schools. A natural reaction of those teachers is to focus on non-teaching duties to balance their lackluster performance in the classroom. The result is that administrators, who are in many cases overwhelmed by their duties and responsibilities, immediately identify potential recruits to the administrative track.

Thus, many potential candidates to the administrative track tend to dislike classroom teaching which leads to the contradictory situation of promoting those who are least passionate about teaching. Moreover, the way the system is structured in terms of formal authority over instructional design and evaluation focuses on the role of the administrative track. Assistant principals and principals are entrusted with evaluating teachers on their teaching. The irony is that the people entrusted with instruction- al leadership chose the leave the teaching track and tend to dislike teaching. Many of those entering the administrative track only have three years of teaching experience, in most cases in subjects different from those they must evaluate, and they are entrusted with evaluating veteran teachers. This leads to very peculiar situations where an administrator with three years of experience teaching an elective must evaluate the teaching of a veteran teacher with fifteen years of experience and several teachers of the year awards. Thus, the incentive structure and the administrative structure leads to a very contradictory situation which can be demoralizing for veteran teachers. Having former lackluster teachers evaluate veteran teachers creates organizational conflict and leads to low morale.

The pay scale is also very different for the two tracks. Administrators receive considerably higher salaries and thus the message that is sent to teachers is that administration is more valuable than teaching. The difference in pay between administration and teaching is not as evident in higher education where many full professors receive considerably higher salaries than many administrators and thus the difference cannot be inherent in the nature of the job (Thelin, 2011). Thus, an outstanding veteran teacher makes less after two decades of teaching than a mediocre assistant principal.

There are many reasons for the wide gap in terms of pay between the two tracks. A structural factor is partly to blame, the role of unions in the negotiation process. Even though teachers' unions have fought for decades for better working conditions and do provide support to their members, one negative structural externality of their centrality in the collective negotiation of teachers' contracts is that they favor contracts that minimize differences between teachers. In other words, the goal is to negotiate uniform pay raises and to play down any differences among the member- ship. Thus, most collective contracts for teachers avoid performance-based incentives and de-emphasize rank among teachers.

One possible track for advancement in the teaching track would be for teachers to seek advanced degrees and receive supplemental pay based on educational achievements. Most districts do have pay supplements for advanced degrees, nevertheless one average the supplements do not reflect the high cost and effort of achieving an advanced degree. The supplement for a master's degree is approximately $2500 per year, for two master's degrees it is $4000, for a doctorate it is $6000. Taking into consideration the high cost of graduate education in the United States, it would take many years for a teacher to recover the investment. The reason for the small supplements is the influence of the unions. Even though many ambitious teachers want to pursue advanced degrees and to improve their skills, union leaders have an interest in achieving improvements in the working conditions of the majority, lowest common denominator, of the membership, and therefore there is an incentive structure to favor raises that apply to all members rather than merit-based incentives such as advanced degree supplemental pay. Differential pay for advanced degrees is perceived by some union leaders to be a divisive issue among the membership. Historically high school teachers were paid higher salaries than middle school and lower schoolteachers in many districts (Goldin, 1999). Many Unions fought against the differences and today pay is the same regardless of level in most districts across the country. This is one example of how Unions tend to promote conformity rather than differences based on merit.

Contextual Factors

Structural issues provide only part of the explanation, contextual factors such as demographic changes and socio-economic trends interact in complex ways with organizational and structural factors. The historically low unemployment rate in the United States at the time of the writing of this study, 3.5 percent, and the even lower unemployment rate in the State of Florida, 2.7 percent, certainly plays an important role in terms of the teacher shortage (*Governor DeSantis Announces Florida's Unemployment Rate Drops to Historic 2.7 Percent as State's Job Creation Skyrockets*, 2022). Low unemployment combined with a booming economy, Florida's economic growth is 6.3 percent, which is higher than the national average. Another important factor is the continued growth of the Hispanic community in the State.

The low unemployment rate plays an important role in in terms of making the recruitment of qualified teachers more challenging. Among many reasons, the availability of well-paid jobs requiring some of the same skills required for the teaching profession, means that school districts must compete for qualified candidates against a vast array of employers. Low unemployment means that recruitment is taking place beyond the traditional disciplinary and professional boundaries. In other words, it is becoming more common for headhunting firms to focus on competencies and skills rather than on specific degrees for recruitment for certain positions.

One case in point is criminal justice. Security agencies including the Federal Bureau of Investigation (FBI) are actively recruiting among teachers («FBI: Special Agent Expert Education/Teaching», 2022). The reason is that teachers have many of the skills required for a successful career as an FBI agent. One position for the FBI is labeled «Special Agent: Education/Teaching Background» and includes the following description «your education experience, whether K-12 or higher, can easily translate to a Special Agent career, where you'll enhance your team by identifying threats and building relationships with communities...» («FBI: Special Agent Expert Education/Teaching», 2022). The starting salary of the position is $78,000 to $153,000. This particular job is based in Tampa, Florida. Even though a career change from teaching to serving as a special agent may seem anomalous yet many of the skills required for teaching overlap with those required to be a special agent. Examples include, *inter alia,* attention to detail, communication skills, leadership skills, and classroom management skills.

Other fields recruit teachers, such as management, hospitality, and real estate. The skills developed by experienced teachers make them great managers, able communicators, and tested leaders. Thus, the low unemployment rate as well as the realization by human resources professionals that teachers have many of the most highly demanded skills, has resulted in large scale poaching of experienced teachers. Low unemployment combined with a relative high inflation rate make it easier to poach teachers to other more lucrative fields.

Poaching also happens from public schools to private schools. In certain regions of the country, due to a strong economy, and the phenomenon of highly skilled digital nomads flocking to highly attractive locations such as

Florida, private schools have seen a sharp upsurge in enrollment (Gibbons, 2022). Examples of this particular phenomenon can be seen in luxury areas of Palm Beach County and Miami-Dade. The upsurge in demand for private schools results in hiring pressure. Elite private schools hire experienced teachers with advanced degrees. Even though salaries tend to be similar to those offered by public schools there are a few important differences. Private schools are not tied to rigid salary schedules and thus have the freedom to negotiate individually with prospective teachers (DeVos, 2022). Public school salary schedules favor equity and thus the gap between the top teachers and the lowest qualified teachers is not large. On the other hand, private schools tend to prefer to hire outstanding teachers with impressive credentials, in particular advanced degrees in the fields they teach. Moreover, private school teachers offer the opportunity for teachers with advanced degrees to focus on teaching the content rather than focusing on the lowest per- forming students and on pedagogy.

Elite private schools tend to have three strikes policies and minimum academic performance requirements and thus the average student is the equivalent of a gifted student in a public-school accelerated program. Therefore, the demographics of elite schools, in addition to financial incentives, and an emphasis on content rather than on process are a few factors facilitating the poaching of experienced public-school teachers by the private education sector. Even though the numbers are small compared to the demand for teachers in the public sector, the effect is magnified by the type of teacher recruited in a time of scarcity of teachers in general. In other words, the best experienced teachers are the ones who are targeted in a time of extreme scarcity. The result is that many public schools are left with a mix of uncertified teachers and rookie teachers.

Strategic Blunders

The previous sections have discussed the role of contextual and structural factors in the current teacher shortage crisis. Nevertheless, a large part of the crisis can be attributed to strategic blunders. The ideological struggle over education has exacerbated the challenges of seeking a consensual policy making process (Kamenetz, 2021). Even though structures determine incentives, there is no doubt that personal agency can also play an important role in the policymaking process. At the macro level the Governor of Florida, Ron DeSantis, has attempted to regulate education through legislative action (Manchester, 2022; «Parental Rights in Education», 2022). He has also attempted to standardize salaries from the top down, a very difficult thing to do because of the decentralized nature of the primary and secondary education systems (Trischler, 2022). Thus, the Governor has tried to pro- vide the resources and to reform the incentive structure to aid lower levels of governments in their efforts to lower turnover and to attract new teachers to the profession.

The challenge is that many principals and superintendents have misinterpreted the guidelines received from the State legislature and others are simply not equipped to implement the re- quired measures. One example is the issue of local referenda for extra funding for education. It has become customary for school districts to propose local referenda to increase the contribution of property owners to provide extra funding for education. The usual reason is to boost teacher salaries in areas with high housing prices.

This type of extraordinary funding can provide an important lifeline to lower turnover by providing funds to raise teacher salaries through supplements of between $5,000 and $10,000 per annum. The funds are not used solely for teachers' salaries, and it is a way for district leaders to seek funds from the public under the cover of helping teachers. Nevertheless, it should be noted that the temporary increase in taxes to fund education must be brough to a vote every three to five years and voters can turn down the request.

Taking into consideration that the minimum teacher salary in the State of Florida is $47,500 and that most supplements pro- vided by local referenda range from $5,000 to $10,000 they play an important role in ameliorating turnover. The problem of strategic blunders is clearly visible in the actions of certain superintendents and principals who ignore the fact that most property owners in Florida tend to be conservative, and due to the extremely high real estate prices in the State they also tend to be older adults («Florida Realtors Market Data», 2022). Therefore, local actions that may antagonize voters jeopardize the passing of referenda. Examples include principals who get involved in highly charged ideological battles over the type of books used in the classroom, superintendents who prioritize infrastructure over teachers, and horror stories of principals who make teachers push trash cans during lunchtime duty.

The point of those local actions is not whether they are legal or not but rather how those actions are strategically counterproductive in seeking the support of local property owners. A district that is replacing the entire fleet of relatively new buses will find it difficult to make the case to voters that it needs money to supplement teachers' salaries. Local stories of principals mistreating teachers with extra duties which are completely unrelated to teaching will under- mine the district's message of seeking to improve the working conditions of teachers. Reasons for strategic blunders by educational leaders range from lack of proper training to insulation from constituents. Many principals were educated at a time when there was no shortage of teachers and were mentored in a top-down leadership style borrowed from the business sector. Other educational leaders simply lack an understanding of how their individual actions support or hamper district and statewide efforts. Sadly, a few rotten apples spoil the entire harvest, and at a time of severe crisis in education, local strategic blunders can have oversized effects on the overall strategy.

Conclusions

The perfect storm is happening due to a vast array of factors such as the decentralized structure of the American education system, the current socio-economic and political context has exacerbated the crisis, and strategic blunders at the local level have foiled many of the strategies developed at the State level to ameliorate the teachers' shortage. Thus, the complex challenges faced by educational leaders require an understanding of the inherent complexity of the problem and the ability to coordinate efforts with a range of stakeholders at the local, state, and federal levels. The effects of the crisis are being felt at the school level and there will be drop in learning indicators at the national level. This is worrying at a time of increasing international competition in which American productivity is lagging many other highly industrialized countries. PISA scores clearly show that American students are not on par with their peers in Asia and Europe, which is surprising taking into consideration that the United States invests more per student than most other OECD member countries.

Several conclusions can be tentatively reached based on the current teachers' shortage crisis. Teaching is not an attractive career for many young Americans not only because of issues of pay but because of the incentive structure of the profession itself (Stern & Johnston, 2013). Moreover, the philosophy of «No Child Left Behind» which focused attention and resources on the bottom 15% of students has not produced the expected results and instead has resulted in abandoning the other 85% of students. American education at the K-12 level is costly and pours most resources on the bottom 15% rather than on most of the student population. A philosophy of student-centered education ironically results in a passive expectation that teachers must solve every conceivable problem to help students achieve unrealistic goals determined by elected officials with no or very little experience in the field of education. A second tentative conclusion is that the sharp separation of a teaching from an administrative track has created a perverse incentive structure for the worst teachers to seek the administrative career and then in turn determine the working conditions of the best teachers. Furthermore, the endogamy of certain school districts in terms of hiring administrators further exacerbates the problem by preventing the influx of new ideas and increasing healthy competition for principalships to attract the most qualified candidates.

Dealing with the current crisis requires more than a local cosmetic superficial fix, it is time for a national dialogue on the very essence of education which includes openly debating the dog mas established by the «No Child

Left Behind» revolution (DeVos, 2022). Even though the task ahead of educational leaders is daunting one, the following section provides a few simple yet potentially impactful recommendation to act and through concerted efforts rejuvenate the educational endeavor at a critical moment in American history.

Recommendations

America has grown through the creative destruction of the free market and through the unbridled spirit of its people. Competition begets excellence and it does not preclude cooperation. Nevertheless, competition cannot be relegated to institutions, but it must infuse the entire enterprise, from the individual student to the whole nation. «No Child Left Behind» corrupted the entire system by distorting the core mission of education and by replacing individual responsibility and healthy competition with passivity and a culture of entitlement. The entire responsibility over the failure or success of the education process was placed on the shoulders of teachers without providing the proper resources to accomplish their mission and with a sharp emphasis on accountability. Accountability for the teacher not for the students.

Thus, change can begin with small individual steps to reestablish individual accountability and a more rational distribution of resources for education. The bottom 15% cannot receive more resources than the top 85%. Moreover, the authority of teachers must be reasserted and the link between results and student effort must be strengthened. It will take a return to the American philosophy of «pulling yourself by the bootstraps» with the proper resources provided by the education system. Results can- not be artificially guaranteed. Educational success is not an entitlement. Education is an active process that requires effort and aptitude from both the student and the teacher.

Healthy competition between and among teachers must be encouraged rather than discouraged. Sclerotic union norms focused on seniority rather than ability and merit must be replaced by a culture of excellence with a clear incentive structure for teachers to improve performance and seek advanced credentials. Salary scales and job titles must support an emphasis on merit rather than on seniority. A related change must be made to education programs at the bachelor's level with a greater balance between case studies focused on public education and private education.

Bibliography

Anderson, J. E. (2006). *Public Policymaking* (6th ed.). Boston: Houghton Mifflin Company.

Bronstein, E. A. (2020). «It isn't in the Curriculum»: World History Teachers' Views on U.S. Presidential Elections. *The Social Studies, 111*(3), 123-132.

Czachor, E. M. (2022). Cardona calls on states to use COVID-19 relief funds to address teacher shortages. *CBS News,* August 21. Retrieved from https://www.cbsnews.com/news/miguel-cardona-covid-relief-funds-teacher-shortage-face-the-nation/

DeSantis, R. (2022). *Governor DeSantis Announces Florida's Unemployment Rate Drops to Historic 2.7 Percent as State's Job Creation Skyrockets.* (2022). Retrieved from Tallahassee https://flgov.com/2022/08/19/governor-desantis-announces-floridas-unemployment-rate-drops-to-historic-2-7-percent-as-states-job-creation-skyrockets/

DeVos, B. (2022). Hostages No More: The Fight for Education Freedom and the Future of the American Child: Center Street.

FBI: Special Agent Expert Education/Teaching. (2022). Retrieved from https://dejobs.org/dallas-tx/special-agent-expert-educationteaching
/2d074bfc3b54483993e9a31bb96a35b3/job/

Florida House of Representatives. (2022. *Parental Rights in Education*, HB 1557 Stat, July 1.

Florida Realtors Market Data. (2022). Retrieved from https://www. floridarealtors.org/newsroom/market-data

Gibbons, P. R. (2022). Florida's private school enrollment eclipses 400,000 students. Retrieved from https://www.reimaginedonline. org/2022/07/floridas-private-school-enrollment-eclipses-400000- students/#:~:text=Private%20school%20enrollment%20grew%20 by%2014.2%25%20this%20year%2C%20compared%20to,1.5% 25%20growth%20for%20public%20schools

Goldin, C. (1999). *A Brief History of Education in the United States.* Historical Paper 119, August, 1-75. Retrieved from https://www.academia.edu/48959462/A_Brief_History_of_Education_in_the_United_States

Hird, John A. M. R., and Matthew Shilvock. (2004). *Controversies in American Public Policy* (3rd ed.). Belmont: Wadsworth.

Kamenetz, A. (2021). Why education was a top voter priority this election. *NPR*. Retrieved from https://www.npr.org/2021/11/04/10521 01647/education-parents-election-virginia-republicans

Kantabutra, S., & Saratun, M. (2013). Sustainable leadership: honeybee practices at Thailand's oldest university. *The International Journal of Educational Management, 27*(4), 356-376.

Manchester, J. (2022). Fight over parental rights in schools reaches fever pitch. *The Hill*. Retrieved from https://thehill.com/homenews/campaign/593937-fight-over-parental-rights-in-schools-reaches-fever-pitch

Index